NEC4: Effective Preparation and Management of the Construction Programme

NEC4: Effective Preparation and Management of the Construction Programme

Tim Willis

Published by Emerald Publishing Limited, Floor 5, Northspring, 21-23 Wellington Street, Leeds LS1 4DL.

ICE Publishing is an imprint of Emerald Publishing Limited

Other ICE Publishing titles:

Early Contractor Involvement
Michael Smith, Matthew Finn, Jon Broome and Catherine Maddox.
ISBN 978-1-83549-897-2
Essentials of Construction Planning and Scheduling
Graham Robertson. ISBN 978-0-72776-592-5
Procurement and Contract Strategies for Construction
Ian Heaphy. ISBN 978-0-72776-371-6

A catalogue record for this book is available from the British Library

ISBN 978-1-83662-481-3

Cover photo: srgktk/iStock.com

Commissioning Editor: Michael Fenton
Content Development Editor: Cathy Sellars
Books Production Lead: Benn Linfield

Typeset by KnowledgeWorks Global Limited
Index by KnowledgeWorks Global Limited
Printed and bound by CPI Group (UK) Ltd, Croydon, CR0 4YY

Contents

Preface

I have been involved with the NEC (New Engineering Contract) family of contracts since the early 2000s' providing training courses and webinars both at home and abroad and accrediting *Project Managers* for the NEC 4 ECC. In the course of that time, I have provided training to and had the privilege of discussing their experience of projects with thousands of users across many sectors. I also practise as a construction lawyer, dealing with both contract advice for projects and disputes in adjudications, the courts and arbitration.

The central purpose of the NEC family of contracts has always been to provide a collaborative basis for those involved in projects to deliver the underlying content as effectively as possible. To achieve that purpose, the NEC family relies on procedures that are recognised as fundamental good project management. Other forms of contract include processes and as a construction lawyer it is often these processes that are the subject of scrutiny in the context of disputes, which centre around adjustments to contract prices or adjustments to dates for completion or other commercial risks. Typically, a dispute is a retrospective examination of the project based upon what information is available at that point in time to analyse liability.

The fact there is a dispute about such adjustments shows that the process was either not well designed or that a party is not following the process.

The NEC contracts seek to achieve an agreed adjustment to prices and completion or key dates, thus precluding such disputes and providing an agreed basis for the future delivery of the project. At the heart of this process is the Accepted Programme, providing an agreed programme of the relative contributions of those involved – be that the client or statutory undertakers or the contractor. The Accepted Programme also provides the basis for predicting the impact of events upon the programmed delivery to allow risk to be managed, contract adjustments to be made and commercial reporting to be up to date.

The Accepted Programme is a central document and yet programmes, and programming more generally, have a poor record in our traditional forms of contract, either playing little or no role in the commercial processes or being fundamentally mistrusted as document on which to place any reliance.

Without an Accepted Programme the NEC ECC contract does not function as envisaged. It is important, therefore, for clients and their contractors to give the preparation and management of this document proper regard.

The purpose in writing this book has been to seek to show what good practice in the preparation and management of the Accepted Programme looks like and the resources that it requires. There would appear to be a need for better understanding of this perhaps because programmes and programming are poorly dealt with in the construction industry. Therefore, contract preparation and management when using the NEC family requires a step-up in performance in this area.

There are too many examples of NEC projects where the programming is deficient and so the book also looks at the implications of failing to properly prepare and manage the Accepted Programme. The consequences as discussed are significant and wide ranging. Not least, anyone who has been involved in trying to resolve disputes will be all too aware of the cost of lawyers and experts pursuing a legal process and the uncertainty of outcome that plagues such disputes. The costs of disputes far outweigh the costs of thinking about programming at the procurement stage and properly resourcing the project management during delivery. The strain imposed by disputes on those involved in the project and the business should also not be underestimated. Good programming makes good business sense and makes for successful projects.

Tim Willis MA, FCIArb, Solicitor Advocate, NEC4 accredited Project Manager and NEC professional.

Acknowledgements

I want to thank the following people for their contributions to this book that have made it possible.

Manoj Bahl of FTI Consulting who originally collaborated with me on the NEC users conference workshop on programming under the NEC4 ECC and who has co-authored Chapters 1 to 4.

John Marshall of Fermata Ltd who contributed the illustrations in Chapter 7 as follows: 7.16 to 7.21.

Dr Michael Fenton, commissioning editor of ICE Publishing at Emerald, who commissioned this book and provided invaluable support and advice during its production.

Thomas Telford Ltd, NEC Contracts and the Institution of Civil Engineers for their support and kind permission to use excerpts from the contracts and illustrations taken from their training courses and user guides to explain some of the central concepts in the NEC4 ECC.

FIDIC and Sweet & Maxwell for their permission to reproduce contract clauses and textbook excerpts.

The Society of Construction Law for permission to quote the SCL Delay Protocol 2nd Edition.

About the author

Tim has practised as partner in law firms, heading up construction and renewables teams and advising clients in the construction and engineering sectors, and as the director of multidiscipline commercial consultants in Hong Kong and the UK. He is an accredited NEC project manager and tutor for ICE training and provides accreditation training to project managers on the NEC4 ECC and training to clients on the NEC family of contracts. Tim has worked in the UK and Asia-Pacific on a broad range of projects and disputes under differing forms of contract and in differing jurisdictions. He is a Fellow of the CIArb, a member of the Technology and Construction Solicitor's Association. Tim is a former chair of the Adjudication Society and on the panel of UK Adjudicators. He originally qualified as a barrister and is a Solicitor Advocate dealing with all forms of dispute resolution. Tim is part of the Construction and Engineering Team at Irwin Mitchell LLP. He is also the owner of Willis Consultancy, providing training, advice and dispute resolution services.

Willis T
ISBN 978-1-83662-481-3
https://doi.org/10.1108/978-1-83662-478-320251002

Chapter 1
Aims of the book

1.1. Aims of the book

Clients and the construction supply chain work in a web of interfaces and dependencies. All the participants plan their contributions from the most strategic level down to the detailed delivery of very specific content or timing of contributions.

The delivery of that content or contributions is often prescribed in a contract and the contract documents impose obligations that seek to manage the delivery of that content or a party's contribution.

The contract might impose a completion date or dates that the client considers important. The contract might impose constraints on how the other party may organise and deliver the content. The contract documents set these dates or constraints at the point it is entered into, but in the context of construction or engineering contracts the underlying content or timing of contributions is subject to change.

The management of change is a central function for clients and the supply chain. Clients and the supply chain have procedures for managing change and employ project managers or contract managers who have responsibilities to manage change. The dates set by a client may need to be adjusted and changes to content or constraints or external events will also impact on the timing of the delivery of content.

The contract often requires that the detailed arrangements are documented in a programme or schedule.

The general use of programmes / schedules and the dedicated function of 'the planner' or 'scheduler' who prepares and analyses programmes is an essential part of the construction and engineering sector and the delivery of complex projects.

The use to which the programmes or schedules are put can be prescribed in the contract or individual contributors may simply produce these documents for their own purposes and share them as they choose to.

The NEC family of contracts use the contract conditions to set procedures to be followed for risk management and to adjust the contractual position of the parties where risk or change impacts the delivery of their respective content or contributions.

The NEC contracts create collaborative processes. Of central importance to risk management is the management of time and the NEC contracts use programmes to manage time and adjust the contract.

By setting a contractual process the contract imposes on the parties a common structure and a minimum standard of procedure to allow the process to function. This has at the heart of it a programme that is subject to continual update and review. It has different functions in the context of the contract.

While clients and contractors may all be familiar with and have planning/scheduling resources, the uses for which they use these resources will differ.

The aim of this book is to identify to both clients and contractors the differing functions and uses of the programme in the NEC family of contracts.

The aim of this book is to place the programme into context in an NEC contract to allow clients and contractors to put in place the procedures and resources to maximise the benefits of using the contract.

The book concentrates on the NEC4 Engineering and Construction Contract (NEC4 ECC), but the book also considers other contracts in the family where detailed programming is required such as the NEC4 Professional Service Contract (NEC4 PSC), the NEC4 Term Service Contract (NEC4 TSC) and the NEC4 Facilities Management Contract (NEC4 FMC).

As will be apparent from the following chapters, programming is central to the NEC4 ECC contract.

The book will also consider the implications of not putting in place and following the procedures. The analysis shows that not programming in the manner envisaged creates significant risk to project delivery and that time spent in understanding the process required and putting it into effect is time well spent.

1.2. Terminology

In the NEC family of contracts, the parties to the contract and those who have specific roles within the contract are referred to in *italicised* font in the contract conditions and identified in the Contract Data Part One or Part Two. For the purposes of discussing specific NEC contracts, that convention is adopted in the text and reference will be made to the parties to the contract, the '*Contractor*' and the '*Client*', and to the '*Project Manager*' and '*Supervisor*'. Where a more general discussion about forms of contract or industry practice is the subject, the italicised terms are not used, and the term employer may be used instead of client (with references made also to other functions variously described as the engineer or architect or contract administrator).

The NEC family of contracts and the naming conventions used is set out in Table 1. Reference is made in this book to other standard forms of contract published for use by the industry. Reference may be made to general families of contracts or specific contracts. A list of the other contracts referred to in this book is contained in Table 4.

Willis T
ISBN 978-1-83662-481-3
https://doi.org/10.1108/978-1-83662-478-320251003

Chapter 2
Introduction

2.1. The unique nature of the construction industry

The construction of buildings or infrastructure is a collaborative process undertaken in a commercial context. We do not think of buildings or infrastructure as 'off the shelf' products. Instead, we construct them to specific requirements, to address specific needs with the resultant product a result of a complex interaction between the procurers and the supply chain.

There is great flexibility in the way that we arrange who will be responsible for the differing aspects of the construction process. Differing procurement practices have developed, and different clients or contexts require that we prioritise different aspects of the procurement to meet their objectives.

2.2. How construction projects are delivered and the NEC family of contracts

One of the objectives in procurement is to provide the basis for organising the delivery of the underlying content. The contracts that are entered into form part of an overall network to achieve these objectives.

The degree of collaboration, and the need to accommodate that as part of a procurement strategy, can vary depending on several factors such as the complexity of the result, the extent to which that has been prescribed, and the number of people who commercially share responsibility for the outcome.

The NEC family of contracts reflect that underlying reality by providing differing contracts to use in differing procurement structures and flexibility within contracts.

The NEC family include a framework contract for procuring work or services, term contracts for procuring the delivery of a service, works contracts, professional services contracts, supply contracts and other contracts that can be used or combined in different ways to procure and integrate delivery of the content of standalone projects. They can also be used under framework contracts to deliver broader procurement objectives across multiple projects.

The family differentiates between complex projects and simpler work and services requirements. Where there are simpler procurement needs, the family has 'short' versions of the differing types of contract – be that work services or supply of goods.[1]

The *Client* must make the strategic decision about how to use or combine the contracts into a given procurement solution. Where short contracts are used, the standard drafting provides the procurement solution without further choices being necessary. Where the full contracts are used, the *Client* must make further selections to determine the contract conditions that will apply. The NEC family makes use of

[1] For a full list of the NEC family of contracts 4th Edition, see Table 1.

differing payment options to deal with pricing and price risk. For the full contracts, the NEC family also make use of secondary options[2] to allow further customisation within each contract if required. The *Client* makes the selection, and the drafting then provides the mechanism for delivering the required content in that context. So, there is considerable flexibility.

The strategic co-ordination of delivery is concerned with what content can be delivered over a period or within an overall budget structure and is based upon the *Client's* overall plan and objectives. Such plans and objectives are bespoke but having decided upon a procurement structure and content to be delivered each individual contract, whether for delivery of work or services, must programme the delivery into a fixed period dictated by the contract.[3]

Depending on where in the procurement structure the particular contract sits, the requirements in terms of programming differ.

The focus of this book is to consider how the works contract that would be used for a stand-alone complex project, the NEC4 ECC, manages the works to be delivered by a completion date in the context of specific requirements as specified by the *Client*. As we shall see, this requires detailed programming placing specific content and interfaces into a specific point in time. The detail and time-bound nature of the delivery gives the programme a structure and purpose that goes beyond strategic planning or aspirations and produces a commercially binding structure to the delivery of the content. It also provides the basis upon which the relative contributions are co-ordinated.

2.3. Requirement for planning from a contractual perspective

In a simple context, where the content to be delivered is well understood by the supply chain and requires little input from others, the planning of work may not be of particular interest to the client. That is reflected in the short forms of contract within the NEC family, whereby the contracts leave the question of whether a programme will be needed open to the *Client*.[4]

This, however, is somewhat of an exception that is appropriate where the underlying content is quite simple but, as we shall see, the underlying ethos behind this suite of contracts is to prioritise good communication, and in a complex project the programme itself plays a significant role in the management and communication processes.

This recognition of the use of the programme as an effective tool for collaboration is a development in which the NEC Engineering and Construction Contract has been at the forefront. Many of our traditional standard forms did not recognise the importance of the programme but, partly because of the experience gained from the use of the NEC Engineering and Construction Contract, a more general desire to see

[2] Secondary Options in the NEC contracts are identified as X or Y clauses that provide standard optional clauses to be selected by a *Client* on a project by project basis to meet the *Client's* needs.

[3] Since the strategic and the detailed programmes are related, there is a strong argument for consistency of approach. The NEC family provides consistency at the project level – for example, the professional service contract has the same programming approach as the works contract.

[4] NEC4 engineering and construction short contract clause 31.2: 'the *Contractor* submits programmes to the *Client* as stated in the Scope'.

more effective collaboration has meant recent thinking has begun to attach greater importance to programmes. If we look at other standard forms of contract published, we see that there is variety in the importance attached to how the contractor will programme the work. Furthermore, some sectors of the industry have attached greater importance to this than others.[5]

The JCT contracts 2016 Editions used for much of building work within the United Kingdom are a good example of a traditional approach that it should be left to the contractor as to how it will complete the work and does not seek to use the programme as a collaborative tool. Instead, this suite sets out that the client's concern is certainty of price and completion date, which is to be achieved contractually by significant transfer of risk to the contractor and leaving it to the contractor to manage the risks transferred or face the consequences of not doing so.

If the contractor does not plan its work properly and is delayed, the client will have its remedy in the form of a claim for losses for late completion.

Such an approach has its proper place, but it is increasingly questioned whether that is an effective approach to procurement. Where the contractor is dependent on the efficient co-ordination of inputs not only from its supply chain but also the client or others outside of its supply chain the need to plan increases as it becomes necessary to co-ordinate the delivery of the underlying content and the contributions of all involved.

There is a further reason why understanding the plan for work is necessary and that is the fact that construction projects are by their nature subject to change. Even where we effectively plan the work, construction contracts must respond to risk that materialises post contract.

In the review of construction contracts carried out by Sir Michael Latham's *Constructing the Team*,[6] traditional forms of contract were seen as adversarial in their approach and that they were a hindrance to the collaborative working processes necessary to deliver complex buildings and infrastructure efficiently and manage risk.

Contrary to this, however, the NEC family was singled out in this review as a form of contract that promoted collaborative processes partly on the basis that it does not seek to make price certainty the primary objective of the client over all others. Instead, the NEC family has flexibility in terms of pricing of content and risk along with the allocation of risk.

The full contracts all contain pricing options ('Main Payment Options'). The options give a range of pricing solutions from lump sum, re-measurement, target cost, cost recoverable to management contracts. For all options there are a sizeable number of 'compensation events' events that have the effect of allocating those risks defined as compensation events to the *Client.*

---

[5] For example, compare the contracts produced by the Joint Contracts Tribunal (JCT) for general building with those of the process engineering industry, such as the Institute of Chemical Engineers, whose contracts are much used in the process engineering sector and contain detail programming requirements reflecting the highly complex and time critical nature of such projects.

[6] Latham M (1994) *Constructing the Team – Final Report of the Joint Government/Industry Review of Procurement and Contractual Arrangements on the United Kingdom Construction Industry.* HMSO, London, UK.

Where content (be that work or services) is to be priced, it is on the assumption that there is clarity about what is to be priced together with the risk that is to be priced. Where that is not practicable because the content or risk is not properly price-able, the pricing options provide alternative solutions.

Where according to the chosen pricing option, content or risk is not to be priced or insured, it is held by the *Client* to be managed collaboratively. To aid that collaborative management the valuation of changes to pricing is based on the concept of 'Defined cost', which is a resource-based valuation based on the cost of the resources plus a fee. This resource-based method of valuation of change seeks to avoid the response to risk or change being driven by pre-determined pricing, concentrating instead on the cost of resources impacted.

In addition, the early warning procedure allows the potential for taking actions to avoid or reduce the effect of risks before they impact the works.

This approach to collaboration and management of risk requires clear project management to understand the potential impact of risks and the possible solutions to avoid or mitigate the impact of risk. The programme, in the form of an Accepted Programme, provides a detailed agreed basis to understand and respond to risk and how it will impact on the parties' relative obligations and contributions and the effect on the work to be done. The programme becomes a central tool in the management of the contract.

2.4. Why plan and why programme works?

From the contractor's perspective, planning and programming is necessary in any context but in complex projects it is even more so, not just to manage its own supply chain but also to manage and respond to risk. For the client, where it holds risk, it can reduce the cost and delay exposure through the efficient management of risk in the early warning procedure and avoid the cost of disputes over adjustments to Prices and the Completion Date through the compensation event procedure.

The programme sits at the heart of this approach by providing an agreed basis for collaboration, and a mechanism for forecasting and assessing the impact of risks that materialise post-contract on future activities to provide for the better management of risk and enable an adjustment of the contract (if required).

The drafting of the contract hard wires this risk management approach into the programming by making the programme, in the form of an 'Accepted Programme', a contract document. This, in turn, imposes contractual obligations, allocates risk, and provides a basis for an agreed adjustment to the Completion Date where risks materialise.

The programme in the form of an Accepted Programme is a dynamic programme constantly updated and revised as the project progresses and events or change impact upon delivery.

Good planning and the Accepted Programme are at the heart of the NEC contracts and serve the interests of both parties.

Willis T
ISBN 978-1-83662-481-3
https://doi.org/10.1108/978-1-83662-478-320251004

Chapter 3
Function of planning and the programme

3.1. Introduction

Programmes are used to achieve several objectives in the context of a construction project. As outlined in the Introduction, the NEC family of contracts place a great focus on the management of the project. In this chapter, we will review how programmes are used to achieve a range of differing objectives in construction and how other standard forms of contract and the engineering construction contract compare in broad terms in delivering those objectives.

3.2. Construction programmes within a contractual context

The word 'programme' is used in many contexts and for many purposes. However, we are concerned with the programme as defined in a contract and here we consider how the programme that the NEC4 ECC requires compares with the other forms of contracts.

This is best done by considering several features that a programme prepared in line with the requirements of the NEC4 ECC ought to possess in comparison with the requirements, if any, of other standard forms of contract.[7]

In this book, reference will be made to programmes generally but in each contract the content and terminology may change, as may the function of the document or documents in question.

The NEC4 ECC refers to an 'Accepted Programme', which is defined as the last programme that has been accepted. It is a dynamic document. Its content is defined but it is not a single document or piece of information but a collection of documents and information that is provided to the *Project Manager* for acceptance.

The programme is essentially the information that the *Contractor* provides, which covers timing and interrelationship between work and other project content and the progress of the works placed in the framework of a completion date or other set dates.

Given the breadth of the information and differing ways in which it is treated in contracts, the word 'programme' can mean differing things in differing contracts. Even the terminology changes from contract to contract.[8]

[7] The contracts chosen are bilateral contracts that assume a single stage procurement of employer design or design and construct. See Table 3.

[8] The Chartered Institute of Building, CIOB Time and Cost Management Contract TCM15 uses terminology such as Working Schedule and Planning Statement to cover such content.

not change its status. The *Client* is not bound by the content of the programme in every detail, but the Accepted Programme may impose certain obligations upon a *Client* and for that reason any programme must be carefully considered.

3.5. Using programmes to establish the effects of change/delay

Another of the uses to which programmes may be put is to assist with the determination of the extent of delay and its effect on the completion date of the project. The obligation to complete by a set completion date is found in most standard forms of contract as, without a completion date, the contractor will be entitled to a reasonable period to complete the works.

The contract will also have a mechanism to help decide adjustment to the completion date if the cause of the delay is one which under the contract drafting or the applicable law entitles the contractor to an extension of time. So, along with the completion date, there will be a mechanism for assessing the extension of time as part of the contract drafting. This permits the completion date to be adjusted so there is a clear reference point from which the contractor can be held liable for any delay to the completion date, or the date as adjusted.

Standard forms of contract all contain such drafting but how the assessment of delay is made varies significantly across contracts. The assessment of delay is often left to the contract administrator.

This may be on an agreed basis following the acceptance of the extension of time as proposed by the contractor. Alternatively, the assessment is made by the contract administrator either in default of such agreement or based upon a unilateral determination process.

Whether the extension is agreed or assessed there must be a basis for such entitlement, and it should not be left just to the whim of the other party or the contract administrator. The contract drafting must provide that basis and in line with this principle the contract drafting will deal with how the extension of time is decided and may do so in greater or lesser detail. The wording used in the contracts varies significantly. The legal jurisdiction will also provide a context in which such determinations are approached.

The question we are concerned with here is to what extent is the programme relevant to the determination of an extension of time?

In most forms of contract, the programme is at best an evidential tool or part of a record that can be used to help determine this question. So, for example, in the FIDIC (2017) Editions,[14] which do require detailed programming and reporting, SC 8.3 states that the basis for an extension of time is that:

> 'The Contractor shall be entitled subject to Sub-Clause 20.2 [Claims to Extension of Time] if and to the extent that completion for the purposes of Sub Clause 10.1 is or will be delayed by any of the following causes.'

Since then, most construction clients using 'traditional contracts' have avoided putting the tendered programme in the contract.

[14] The FIDIC Red, Yellow and Silver Books (2017) Editions and Red Book 2nd Edition (2017) (reprinted 2022 with amendments) and FIDIC EPC/Turnkey Contract 2nd Edition (2017) Silver Book (reprinted 2022 with amendments).

Willis T
ISBN 978-1-83662-481-3
https://doi.org/10.1108/978-1-83662-478-320251004

Chapter 3
Function of planning and the programme

3.1. Introduction

Programmes are used to achieve several objectives in the context of a construction project. As outlined in the Introduction, the NEC family of contracts place a great focus on the management of the project. In this chapter, we will review how programmes are used to achieve a range of differing objectives in construction and how other standard forms of contract and the engineering construction contract compare in broad terms in delivering those objectives.

3.2. Construction programmes within a contractual context

The word 'programme' is used in many contexts and for many purposes. However, we are concerned with the programme as defined in a contract and here we consider how the programme that the NEC4 ECC requires compares with the other forms of contracts.

This is best done by considering several features that a programme prepared in line with the requirements of the NEC4 ECC ought to possess in comparison with the requirements, if any, of other standard forms of contract.[7]

In this book, reference will be made to programmes generally but in each contract the content and terminology may change, as may the function of the document or documents in question.

The NEC4 ECC refers to an 'Accepted Programme', which is defined as the last programme that has been accepted. It is a dynamic document. Its content is defined but it is not a single document or piece of information but a collection of documents and information that is provided to the *Project Manager* for acceptance.

The programme is essentially the information that the *Contractor* provides, which covers timing and interrelationship between work and other project content and the progress of the works placed in the framework of a completion date or other set dates.

Given the breadth of the information and differing ways in which it is treated in contracts, the word 'programme' can mean differing things in differing contracts. Even the terminology changes from contract to contract.[8]

[7] The contracts chosen are bilateral contracts that assume a single stage procurement of employer design or design and construct. See Table 3.

[8] The Chartered Institute of Building, CIOB Time and Cost Management Contract TCM15 uses terminology such as Working Schedule and Planning Statement to cover such content.

However, it is also true to say that there has been a development of the treatment of programmes in many of the standard forms considered. The FIDIC forms of contract and the IChemE forms now exhibit many of the features the NEC family of contracts contain. It is in general building work, within which the JCT suite of contracts are typically used, that there has, however, been the least development. The CIOB Time and Cost Management Contract by contrast has taken time management further than any other standard form of contract but its use has not been adopted by the market.

This greater convergence we have seen in the treatment of programmes is helpful in that it means that the industry will become better at responding to the challenge of producing and submitting detailed programmes. The NEC family has been in the vanguard of these developments.

Despite such developments, it is still the case that on many projects detailed programming information is not a requirement within the contract, neither in some projects are the full contractual requirements met.

3.3. Using programmes to monitor progress

Depending on the level of detail that a programme must contain and the frequency with which it must be updated, the programme can provide a record of progress. As we shall see in Chapter 5, the Accepted Programme can provide that record if it is updated in the manner envisaged.

Some standard forms of contract approach the monitoring and reporting of progress as a distinct function and place great emphasis on the reporting of progress without directly linking that information as part of the programming content.

For example, the FIDIC forms of contract deal with the programme and progress reporting as separate obligations. The emphasis in FIDIC is on ensuring that the progress reporting is full in content and provided in a timely manner.

3.4. Using programmes to control progress

A related but different focus of the programme in construction contracts is the extent to which it commits a contractor to progress at a specific rate.

The focus in that case is upon imposing potentially binding obligations on a contractor with the intention of being able to enforce the obligation to meet a completion date and potentially intervene if it appears the completion date will not be met owing to some failure of the contractor to progress.

Generally, there is no remedy outside of the contract for an employer who is simply concerned that the contractor is going slowly. Instead, any right to damages does not usually crystalize until the contractor has failed to meet the completion date. Rights to terminate a contract might arise in the case of very bad failure to carry out work but these are of limited use to an employer seeking to preserve a completion date. Added to this is the fact that invariably terminating a contract creates delay and inevitably disputes.

Some forms of contract do, however, seek to make a tighter link between progress and the achievement of the completion date.[9] This clause gives the project manager the power to notify the contractor if the

[9] See, for example, clause 13.5 of the Institution of Chemical Engineers (2013) *IChemE Red Book Fifth Edition*.

progress being made is likely to prejudice its ability to meet the completion date and it is owing to a cause for which the contractor is responsible. Following such notification, the contractor is required to remedy the delay at its own cost.

The FIDIC forms of contract 2nd Editions of the Red, Yellow and Silver Books also impose contractual liability on the contractor for failure to progress works in accordance with the programme. The FIDIC forms go further than the IChemE forms in providing an express remedy. Under clause 8.7, where progress has fallen (or will fall) behind the Programme, the engineer may instruct the contractor to submit a revised programme setting out the revised methods proposed to expedite progress. Unless the engineer states otherwise, the contractor must adopt these revised methods including increases to the working hours, personnel, or resources at the contractor's risk. If any additional costs are caused to the employer, the employer is entitled to these in addition to delay damages.

By contrast, while the NEC family contemplates the possibility of the work being expedited this can only be done through the contractor's own volition (to avoid damages for missing the Completion Date) or through agreed 'acceleration measures'. This approach implicitly reflects an analysis that the contractor even if in delay due to its own fault cannot be asked to speed up.

The closest the NEC family gets to tying a contractor to liability for failure to progress beyond completing the works on or before the Completion Date is in the use of Key Dates. These are optional in that it is not on every project a *Client* will need to include them.[10] Once included, however, they expose the *Contractor* to potential liability to meet Key Dates and, if it then fails to do so, the *Contractor* is exposed to the additional costs of work to be carried out by others (or the *Client*).

Like the Completion Date, however, there exists no remedy to force the *Contractor's* pace if it appears that the Key Dates will be missed, and they can only be changed by agreement or adjusted to a later date through the compensation event process or by instruction.[11]

It has been suggested that a remedy that might be available to a client who considers that the *Contractor* is progressing too slowly is for the *Client* to remove scope from the *Contractor* to have it carried out by itself or others. It could then re-charge that cost to the *Contractor*. That is not, however, supported by the drafting and would quite probably be a breach of contract.[12]

So far, we have considered to what extent the programme binds a contractor to progress at a specific rate. The other consideration that is sometimes raised is the extent to which a client is bound to allow the contractor to progress in accordance with a programme or the consequences if it is no longer possible to progress in accordance with it. Such concerns raise the question of whether the programme should be given a formal contractual status.[13] As we shall see, the *Client* has a choice in the NEC4 ECC as to how the programme comes into the contract but by whatever route it becomes a contract document that does

[10] See para 4.55 for a detailed explanation of Key Dates.

[11] There is a suggestion that under clause 12.3 the *Project Manager* could instruct a change to a Key Date to an earlier date than that originally set in the Contract Data or adjusted.

[12] *Van Oord UK (Limited)* v. *Dragados UK Limited [2022]* CSOH 30.

[13] A contentious issue for clients, however, is whether to include a tendered programme in the contract. In *Water Authority* v. *Sir Alfred McAlpine Ltd [1985]*, a contractor successfully claimed against a client when the sequence and methods in the tendered programme, which the client had included in the contract, proved to be impossible.

not change its status. The *Client* is not bound by the content of the programme in every detail, but the Accepted Programme may impose certain obligations upon a *Client* and for that reason any programme must be carefully considered.

3.5. Using programmes to establish the effects of change/delay

Another of the uses to which programmes may be put is to assist with the determination of the extent of delay and its effect on the completion date of the project. The obligation to complete by a set completion date is found in most standard forms of contract as, without a completion date, the contractor will be entitled to a reasonable period to complete the works.

The contract will also have a mechanism to help decide adjustment to the completion date if the cause of the delay is one which under the contract drafting or the applicable law entitles the contractor to an extension of time. So, along with the completion date, there will be a mechanism for assessing the extension of time as part of the contract drafting. This permits the completion date to be adjusted so there is a clear reference point from which the contractor can be held liable for any delay to the completion date, or the date as adjusted.

Standard forms of contract all contain such drafting but how the assessment of delay is made varies significantly across contracts. The assessment of delay is often left to the contract administrator.

This may be on an agreed basis following the acceptance of the extension of time as proposed by the contractor. Alternatively, the assessment is made by the contract administrator either in default of such agreement or based upon a unilateral determination process.

Whether the extension is agreed or assessed there must be a basis for such entitlement, and it should not be left just to the whim of the other party or the contract administrator. The contract drafting must provide that basis and in line with this principle the contract drafting will deal with how the extension of time is decided and may do so in greater or lesser detail. The wording used in the contracts varies significantly. The legal jurisdiction will also provide a context in which such determinations are approached.

The question we are concerned with here is to what extent is the programme relevant to the determination of an extension of time?

In most forms of contract, the programme is at best an evidential tool or part of a record that can be used to help determine this question. So, for example, in the FIDIC (2017) Editions,[14] which do require detailed programming and reporting, SC 8.3 states that the basis for an extension of time is that:

> 'The Contractor shall be entitled subject to Sub-Clause 20.2 [Claims to Extension of Time] if and to the extent that completion for the purposes of Sub Clause 10.1 is or will be delayed by any of the following causes.'

Since then, most construction clients using 'traditional contracts' have avoided putting the tendered programme in the contract.

[14] The FIDIC Red, Yellow and Silver Books (2017) Editions and Red Book 2nd Edition (2017) (reprinted 2022 with amendments) and FIDIC EPC/Turnkey Contract 2nd Edition (2017) Silver Book (reprinted 2022 with amendments).

The JCT (Joint Contracts Tribunal) SBC 2016 Edition does not require detailed programming and the basis for an extension of time is that if a delaying event is a Relevant Event (as prescribed), and completion of the Works or any Section thereof has been or is likely to be delayed beyond the Completion Date for the Works or a Section thereof, then:

> '. . . the Employer shall give an extension of time by fixing such later date as the Completion Date for the Works or a Section as he then estimates to be fair and reasonable.' (clause 2.28.1)

In neither case is the programme explicitly referred to as part of the assessment.

By contrast, NEC4 ECC refers explicitly to the Accepted Programme current at the dividing date to determine an extension of time. Such an extension of time is required to be assessed as:

> '. . . the length of time that due to the compensation event planned Completion is later than planned Completion as shown on the Accepted Programme current at the dividing date.' (clause 63.5)

This gives the programme in the form of the Accepted Programme a vital role in the analysis and determination of delay and extensions of time under the NEC4 ECC form of contract.

The NEC4 ECC is almost unique in this respect.[15]

3.6. Comparison of the use of the Programme in published standard forms of contract

A table comparing differing requirements of the programme in some of the standard forms of contract commonly used is set out at Table 2.

3.7. The NEC4 ECC Accepted Programme's unique features

We have seen that the requirements in terms of the preparation and updating of programmes vary across differing forms of contract. The NEC4 ECC is not unique in requiring a programme, but it has more features that are required, as well as addressing specific aspects of the programme, which set it apart from the other forms.

First, there is the detail required in terms of content. The two forms that come closest in terms of detail are the new FIDIC contracts and the IChemE forms. As we have seen, others such as JCT set out no standard requirements.

The FIDIC forms of contracts also refer to progress reports that have a quasi-contractual status to achieve and address much of the detail that is contained in the NEC programme.

It is fair to say, therefore, that the NEC4 ECC programme provides the most detailed provisions of the commonly used contracts and its concentration on the programming of the work, together with the requirement to show a planned completion date, hold it apart from other standardised forms of contract.

[15] The CIOB Time and Cost Management Conditions of contract also take such an approach. See clause 51.

In addition, the direct linkage to withholding of part of the Price for Work Done to Date and the submission of the first programme is unique to the NEC4 ECC.

Further, no other contract makes a direct link between the programme and the assessment of the extension of time.[16]

In terms of risk allocation, the programme has a greater role within NEC4 ECC than in other contracts. Contracts will often deal with risks arising from the failure of the employer (or others on its behalf) to provide something necessary for the contractor to carry out its work. It is only the NEC4 ECC which expressly links a resultant entitlement in the form of compensation events directly to dates identified in the Accepted Programme. Three significant compensation events arise by reference to what is shown on the Accepted Programme.[17]

The NEC4 ECC is the only standard form that takes delay to a planned completion date as opposed to the Completion Date or a delay to the Completion Date as the reference point for an extension of time. The gives rise to a principle referred to as 'terminal float', which provides a measure of the period of time between the planned completion date and the Completion Date. The ownership of any period of terminal float present is further defined within the NEC4 ECC.[18]

3.8. Other contracts within the NEC family

Programmes are not unique to the NEC4 ECC contract. The Professional Services Contract (NEC4 PSC) contract, the Term Service Contract (NEC4 TSC) and the Facilities Management Contract (NEC4 FMC) also have detailed programme requirements. These are again used in the context of the compensation event procedure to adjust the Completion Date in the NEC4 PSC and the Task Completion Date or Project Order Completion Date in the NEC4 TSC and NEC4 FM contracts.

The use of the programme in a professional service context is a means of achieving greater integration and control over the underlying services delivery. The compensation event procedure links the programme as part of the risk allocation as it also does within the NEC4 ECC.

In order to properly administer the NEC4 PSC contract, consultants must be prepared to properly plan and programme the delivery of the services and adjust the Completion Date as necessary or face liability for delay. Given that requirement, much of the guidance given here in the context of the works contracts is also of relevance for *Consultants* preparing and manging the programme when using the NEC4 PSC contract.

Similarly, Task Orders and Project Orders under the NEC4 TSC and NEC4 FMC contracts have a completion date and the contract requires that there is a Task Order or Project Order programme. This is again linked to the risk allocation and the adjustment of the Task Completion Date or Project Order Completion Date through compensation events and the concept of planned Task Completion or planned Project Order Completion.

[16] The *CIOB Time and Cost Management Conditions* (2015) refer to a predictive extension of time based upon the Working Schedule (programme) but this is subject to review based upon actual Progress Records. See clause 51.3.

[17] The compensation events are discussed in more detail in Chapter 7.

[18] The concept and treatment of the terminal float is discussed in greater detail in Chapter 7.

The TSC and FMC contracts also have a Plan that covers the content to be provided across the whole term of the contract. The Plan is not a programme organised around a completion date, but nonetheless it is a contract document. Failure to provide 'something in accordance with the Accepted Plan' is a 'Defect' or 'Service Failure', that must be rectified at the *Contractor's* cost and so, the Plan has contractual force with liability for failure to provide the Service in accordance with the Plan.[19]

3.9. NEC4 contracts with less demanding programming requirements

The NEC family of contracts include contracts for high complexity such as the NEC4 ECC and also 'short contracts' where the content/nature of the relationship is less complex.[20]

The selection of a short contract is appropriate where the cost of resourcing the management procedures in the full contracts is not justified by the content of the contract and the risks associated with delivery of content do not need close management.

Where the content can be delivered without the need for significant collaboration with the *Client* or *Others*, and the risks associated with delivery are limited, the cost of the full management procedures and the appointment of a dedicated *Project Manager* or *Service Manager* (when compared with the procurement needs) may be disproportionate or the processes unnecessary.

The short contracts accordingly provide much of the core process but with a less prescriptive approach. This is achieved in several ways.

- The NEC4 ECSC (Engineering and Construction Short Contract), as with the other short contracts, are administered directly by the *Client* and the other party.
- The programming requirements, if any, are left to the *Client* to specify within its Scope.
- There is no direct linkage between the programme and the method of adjusting the Completion Date.
- In addition, there are no compensation events that refer to the programme.

The NEC4 PSSC (Professional Service Short Contract) follows the same approach.

[19] NEC4 TSC clauses 11.2. (5).

[20] Most contracts in the family have a short version. These are identified with the letters SC after the relevant type of contract. For example, NEC 4 PSCSC refers to the Professional Services Short Contract.

Willis T
ISBN 978-1-83662-481-3
https://doi.org/10.1108/978-1-83662-478-320251005

Chapter 4
Preparing the programme

4.1. Introduction

This chapter looks at the how both the *Client* and the *Contractor* are involved in the programming that will be required for a project. Both have an important role to play in ensuring that the objectives of the contract drafting are met.

4.2. The importance of proper planning

The NEC4 ECC contract places the responsibility on the *Contractor* to prepare the programme. Notwithstanding this it is important that the *Client*, when preparing the documents for a tender, considers the content and form of the programme carefully.

In particular, in order for the *Contractor* to plan its work the *Client* is required to consider the relationship between the work that the *Contractor* is to do and how that will be linked to or be dependent upon the work that the *Client* or Others[21] will be carrying out. Furthermore, consideration is also required if information or other things are to be provided by the *Client* or Others and, if so, when that will be provided.

The contract assumes that the *Client* makes the *Contractor* aware of these matters in the Scope. It is in this respect of that it follows that the *Client* is not just the recipient of the programme but also determines the content that the *Contractor* is to include in the programme. If the *Client* does not consider such matters, the drafting leaves it open for the *Contractor* to programme based upon the information that has been made available to it in the Scope, or that which has been subsequently agreed. If agreement is not reached, then the contract is silent on the consequences and resolving such issues is potentially complex.

The importance of the *Client* considering such matters is further set out in the detailed guidance, Preparing an Engineering Construction Contract Volume 2 of the NEC 4 User Guide, for the contract.

Before considering the detail, we must first consider the terminology.

4.3. What is the programme?

The NEC4 ECC refers to a programme, or the Accepted Programme, as follows:

> 11.2 (1) The Accepted Programme is the programme identified in the Contract Data or is the latest programme accepted by the *Project Manager*. The latest programme accepted by the *Project Manager* supersedes previous Accepted Programmes.
>
> 31.1 If a programme is not identified in the Contract Data, the *Contractor* submits a first programme to the *Project Manager* for acceptance within the period stated in the Contract Data.

[21] See clause 11.2(12) Others and page 33 further on.

From the drafting, we can see the contract leaves it open for the *Client* to stipulate if the programme is to be provided by the *Contractor* in response to the tender (as part of the Contract Data Part Two) or, alternatively, a timescale in which the programme must be submitted post-contract. That timescale is identified by the *Client* in the Contract Data Part One (as completed by the *Client*).

Thus, responsibility is placed on the *Contractor* to produce the first programme, but then, as we shall see, the *Contractor* also has the responsibility to revise the programme.

If the programme is submitted as part of the tender process, then the initial control of the content of the programme must be through the tender evaluation procedure. Any revised programme will be subject to review and either acceptance or non-acceptance by the *Project Manager*. Where no programme was required at the tender stage, the contract drafting also gives a special status to the 'first programme' to provide some additional control over the adequacy of the programme submitted post contract.

4.4. The first programme

The significance of the first programme is apparent by reference to clause 50.5 of the contract that enables the *Project Manager* to retain 25% of the Price for Work Done to Date in each assessment until such time that the *Contractor* has submitted a first programme for acceptance (detailing the information which the contract requires). This is the only circumstance within NEC4 ECC that allows the *Project Manager* to withhold monies from the *Contractor* owing to failure to comply with the programming requirements and it reinforces the importance placed on the planning process (and the use of a programme as a project management tool).

This is, however, only applicable to the first programme, and once the first programme is submitted containing the information that the contract requires, then this provision ceases to apply.

The drafting is therefore limited.

From a *Client's* perspective, on the basis that any programme included within the Contract Data becomes the first accepted programme, it is fundamental that a thorough programme is prepared by the *Contractor* at tender stage and included within the *Contractor's* tender submission.

Doing so demonstrates that the *Contractor* understands the project[22] and that the works can be completed within the necessary timeframe and in accordance with the price submitted. This also provides assurances to the *Project Manager* that the *Contractor* has the competency to prepare a robust programme (and use the planning software).

For this reason, the programme submitted by a tenderer ought to be a significant factor in the evaluation of a tender submission.

Conversely, the *Contractor* is advised to consider the practical aspects (such as logic links, durations and the underlying assumptions) of any programme submitted at tender stage on the basis that at the point of any subsequent revision to the programme, the grounds for not accepting a programme (as set out by way of clause 31.3) include failure to include the detail required by the contract and that, as we will see,

[22] Although conversely it could equally be argued that this would only really come following appointment.

require a level of detail that would not naturally apply at this early stage of a project if the programming was done on a more traditional basis.

From both *Client's* and *Contractor's* perspective it is important to consider the detailed requirements of the programme.[23]

4.5. Requirements in relation to the programme within NEC4 ECC

Clause 31.2 provides a comprehensive list of the requirements a programme submitted by the *Contractor* ought to include for it to be capable of acceptance.

This includes the following information:

The programme

31.2 The *Contractor* shows on each programme submitted for acceptance

- the *starting date*, *access dates*, Key Dates and Completion Date,
- planned Completion,
- the order and timing of the operations which the *Contractor* plans to do in order to provide the Works,
- the order and timing of the work of the *Client* and Others as last agreed with them by the *Contractor* or, if not so agreed, as stated in the Scope,
- the dates when the *Contractor* plans to meet each Condition stated for the Key Dates and to complete other work needed to allow the Client and Others to do their work,
- provisions for:

  – float,

  – time risk allowances,

  – health and safety requirements and

  – the procedures set out in the contract,
- the dates when, in order to Provide the Works in accordance with the programme, the *Contractor* will need:

  – access to a part of the Site if later than its *access date*,

  – acceptances,

  – Plant and Materials and other things to be provided by the *Client* and

  – information from Others,
- for each operation, a statement of how the *Contractor* plans to do the work identifying the principal Equipment and other resources which will be used and
- other information which the Scope requires the *Contractor* to show on a programme submitted for acceptance.

A programme issued for acceptance is in the form stated in the Scope.

[23] There is no express requirement for the programme identified in the Contract Data to be compliant with clause 31.2, but a pre-contract programme that does not comply with the requirements of the contract will be rejected when it is revised and submitted post contract.

The clause identifies 17 separate requirements that must be shown on the programme. The programme may be a collection of documents. In addition, the contract envisages that further content or requirements as to the form or content of the programme may be defined in the Scope.

4.5.1 The relevance of the Scope

The contract sets out the requirements for the programme in clause 31.2 but, as mentioned above, the Scope will also play an important role.

As we will see, there is much cross-referencing to the Scope within the contract and in the list of compensation events. This is done to provide reference points for identifying/managing risk and also to identify the content that the *Contractor* must programme against.

In addition to what must be captured in the programme it invites the *Client* to specify the specific requirements the *Client* may have in relation to how the programme is prepared and presented.

The model Scope structure set out within the NEC4 Users Guide[24] (as set out below) suggests a specific section of the Scope dealing with the programme is used and sets out the issues to be considered.

S 500 Programme

The ECC includes detailed programme requirements and procedures. It may be necessary to set out specific *Client* requirements.

Checklist	Explanation
S 505 Programme requirements ECC 31.2	State requirements for the format of the programme, including the use of specific software (if necessary) and the requirement for hard or electronic copies. State any requirement for the programme to be produced in levels (for example summary level to detail level).
ECC 31.3	State any information additional to the requirements of clause 31.2 that the *Contractor* is to show on the programme. This may include dates for submission of designs and samples, dates for information or actions by the *Client* and *Project Manager*, and the timing of any test and inspection.
S 510 Methodology statement	Particular requirements for methodology statements, including any specific requirements for the format of resource information.

[24] *NEC4 Users Guide: Preparing an Engineering and Construction Contract Volume 2* (June 2017, with amendments October 2020) Thomas Telford Limited, UK: p. 66. See: https://www.neccontract.com/products/contracts/nec4/user-guides-and-flow-charts/nec4-preparing-an-engineering-and-construction-contract#:~:text=For%20users%20seeking%20to%20prepare%20their

Checklist	Explanation
S 515 Work of the *Client* and Others ECC 25.1 ECC 60.1(5)	Detail the order and timing of the work of the *Client* and *Others* to be included in the programme and information to be provided. Refer as necessary to sections S 905 and S 910.
S 520 Information required	A schedule of information to be provided, who it is to be provided by and the date by which it is to be provided.
S 525 Revised Programme	State any specific requirements for the submission of revised programmes such as an explanation of changes.

Source: © Copyright nec 2025

The guidance provides a starting point for the *Client* to consider the areas identified and picks out those aspects of the drafting that the *Client* is invited to consider in the Scope. It covers a combination of matters some that relate to the form of the programme, such as software and access to the programme or how it is presented. It also refers to matters that will supplement the content to be included in such a programme. Having considered the starting point of the standard requirements on each project, the *Client* should consider whether these additional requirements should be requested in the Scope. Before considering what additional requirements might be needed the starting point is to consider the elements already required in the conditions and defined at clause 31.2. Each of these is considered in turn below.

4.5.1.1 Programme requirements under clause 31.2

Clause 31.2 lists first a number of dates that have been defined by the *Client* in the Contract Data and are best represented in chorological order. See Figure 4.1.

4.5.2 *Starting date*

This is identified within the Contract Data Part One. It relates to the date upon which the *Contractor* assumes the risks stated in the contract (as set out at clause 81.1) and, consequently, provides the necessary insurances against them (in accordance with clause 83.3). It is also the date from which the payment process commences in line with clause 50.1.

In terms of practicality, therefore, it would make sense that the *starting date* would be before the *access date*. This is on the basis that if the *starting date* was *after* the *access date*, there would be a period between the two dates during which the contractor would not be at risk for any damage it may cause nor would it have the necessary insurances against that damage. In addition, the payment provisions under clauses 50 and 51 would not have a reference point during this same period.[25]

[25] The Contract Date is not referred to in the programme but is the date when the contract came into existence and so due allowance for the procurement process should be taken into account when setting the *starting date*.

Figure 4.1 Defined dates to be included under clause 31.2 (Source: © Copyright nec 2025)

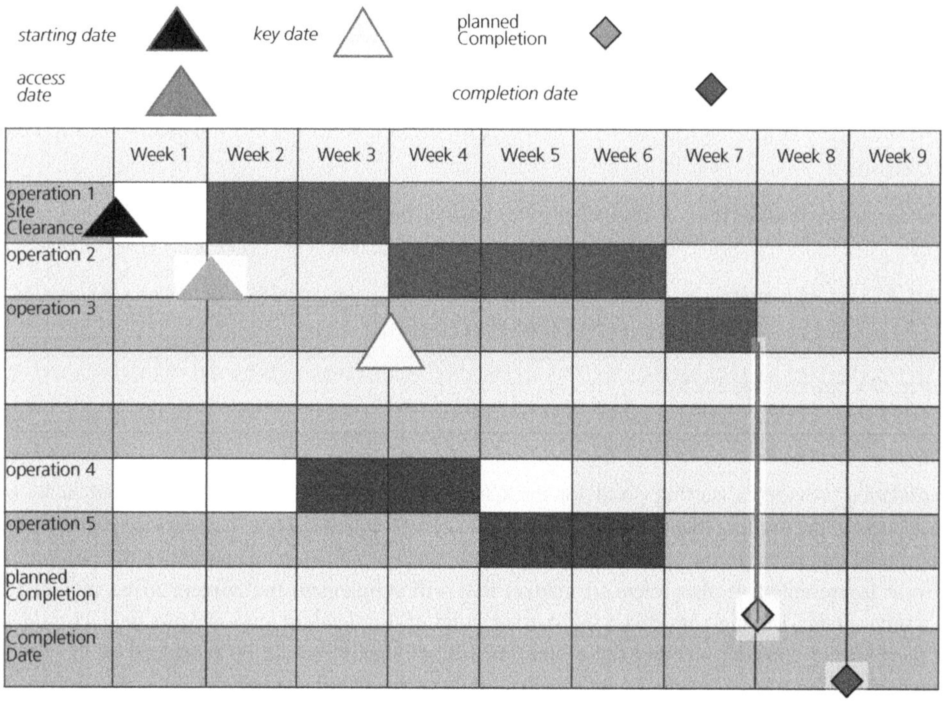

4.5.3 *Access dates*

This is also identified within the Contract Data Part One and reflects the dates upon which the *Contractor* is due to gain access to specific parts (or all) the site. The *Contractor* is entitled to start work on site on the earliest *access date*, albeit not before (in line with clause 30.1).

The *Client* is required to provide access to and use of each part of the Site by the later of its *access date* and the date for access shown on the Accepted Programme. If it fails to do so, in line with clauses 33.1 and 60.1(2), a compensation event will arise.

4.5.4 Completion Date

The Completion Date is a defined term in clause 11.2(3) and is the date stated in the Contract Data unless changed in accordance with the contract. It is the point in time from which delay damages would accrue.

Under 'clause 30.1' The *Contractor* does the work so that Completion is on or before the Completion Date.

This date is fixed and will only move to a later date on account of implemented compensation events. 'Completion' is a defined term.

> 11.2 (2) Completion is when the *Contractor* has
>
> - done all the work which the Scope states is to be done by the Completion Date and
> - corrected notified Defects which would have prevented the *Client* from using the works or Others from doing their work.
>
> If the work which the *Contractor* is to do by the Completion Date is not stated in the Scope, Completion is when the *Contractor* has done all the work necessary for the *Client* to use the works and for Others to do their work.

The Completion Date is fixed but the *Contractor* has in programming its work to consider the definition of Completion. This may require programming against a very clear definition provided by the *Client* in the Scope or against the much less prescriptive concept referred to in the definition of what is necessary to be done for the *Client* to use the Works or Others to do their Work.

4.5.5 Key Dates

The use of Key Dates requires a *Contractor* to meet certain conditions identified in the Contract Data Part 1 ('the *condition*') by a certain date identified in the Contract Data. A Key Date is defined in clause 11.2(11).

The Key Dates require the *Contractor* to give priority for the need for other work to be carried out by the *Client* or Others and to meet the *condition* stated in the Contract Data by that date.

If *Contractor* fails to meet the *condition* stated by the Key Date it is in breach of contract under clause 30.3. Unlike other potential breaches of contract the contract has a specific remedy under 25.3 for the *Client* to recover specific cost arising from that failure. This is the *Client's* only remedy. The *Contractor* is exposed to a commercial risk but unlike delay damages, which if Secondary Option X7 Delay damages applies are defined in the Contract Data Part 1, these costs are not pre-defined.

For this reason, the Key Dates require managing in the same manner as the Completion Date. Key Dates must be adjusted if owing to a compensation event the *Contractor* is unable to meet the *condition* required by the planned date.

4.5.6 Sectional Completion Dates

These will be defined if Secondary Option X5 Sectional Completion is chosen.

Sectional completion is usually used where within the project there are discrete sections that can be completed and handed over to the *Client* to start using once complete. Once a section has been completed, the *Contractor's* exposure to delay damages is reduced.

The *Client* must identify each Section of the *works* and each Section has its own Completion Date and its own planned completion date.

Sectional Completion creates several Completion Dates in addition to the Completion Date for the whole of the *works*.

X5 requires that each reference in the contract to Completion, the *works*, and the Completion Date applies to each Section of the *works*. A reference to the whole of the works is a reference to the last date that is separately identified in the Contract Data.

Much of the following discussion in relation to programming applies equally to each Section.

4.5.7 Planned Completion

The Completion Date and planned Completion are separate items. Each must be shown on the programme.

The requirement is for the *Contractor* to show the date upon which it plans for the works to be completed in accordance with the definition of Completion. This will be determined by the Scope and whether the work to be done to achieve Completion has been specifically dealt with in the Scope. The *Contractor* can but is not required to programme for any work that could be carried out after Completion, but that work is not part of planned Completion. Post completion, the commercial position changes in that delay damages are no longer deductible but also the programming requirements cease on completion of the whole of the works.

The *Contractor* will show planned Completion in its first programme, but this date will move according to the factual circumstances and could be on or before or after the Completion Date. Under clause 30.1, the *Contractor* is obliged to do the work so it achieves Completion on or before the Completion Date. If it completes later than the Completion Date, it will be liable for delay. As we will see, because of the link between assessing adjustments to the Completion Date by reference to 'planned Completion', the *Contractor* whose programmes shows it finishing early owns the 'terminal float'.

Accordingly, Planned Completion denotes when the *Contractor* 'plans' to complete the conditions that would meet the definition of Completion. As such it ought to be properly (logic) linked into the programme and denote the correct sequencing for completion.

On the basis that this is a forecast date this can, therefore, move dependent on the status of the works and whether the *Contractor* is ahead or behind in terms of progress achieved. The planned Completion might move and not have any effect on the Completion Date. If the planned Completion moves to an earlier date it will not move the Completion Date to an earlier Date, and if it moves to a later date than the previous planned completion then the Completion Date will only move if the compensation event mechanism permits the Completion Date to be adjusted.

The contract then goes into more detail about what the Contractor must consider and show in the programme that gives the planned Completion.

4.5.8 The order and timing of the operations that the *Contractor* plans to do in order to provide the works

'Operations' is not a defined term but would include those things necessary in order 'To Provide the Works', which is a defined term.

'11.1(15) To Provide the Works means to do the work necessary to complete the works in accordance with the contract and all incidental work, services and actions which the contract requires.'

The wording encompasses more than just physical work and includes services and processes. There is an express reference in clause 31.2 to allowing for 'the procedures' in the contract. See below.

As we shall see below, the operations also have to be properly linked and the assumptions in terms or resourcing and risk also must be shown.

The contract does not expressly say that the programme must show a critical path but it is implicit in the drafting that this is what is required.

The order and timing of the work is demonstrated by way of the logic links within a programme, which in turn sets out the sequencing of the works and generates the resultant dates.

Whilst the calculation of the dates (and the float) is performed by the planning software itself, the basis for this is the information entered into a programme (namely the activities, the activity durations and the logic links). This is a manual exercise performed by the planner. The quality of the programme is entirely governed by the quality of the information input and the quality of the planner.

The programme must incorporate all the operations required, including adding sensible and realistic logic links to create a CPM (Critical-path method) programme tool capable of measuring progress and change against.

In addition to showing the operations the *Contractor* plans to do, the contract also refers to float and time risk allowances being shown on the programme and also for each operation, a statement of how the *Contractor* plans to do the work identifying the principal Equipment and resources the *Contractor* will use.

The concepts of float and time risk allowances are used in a specific way in the NEC4 ECC and it is important to put them into context.

The contract conditions refer to the requirement to show 'float' and 'time risk allowances' but we also have to consider other terminology that is derived from the drafting or used by planners. In particular, the terms 'terminal float' and 'free float'.

There is no express reference to 'free float' or 'terminal float'.

The contract does, however, create a concept of terminal float and rules as to how it is to be treated. It is one of the most important aspects of the way that risk is allocated.

4.5.9 Terminal float

This can most simply be defined as the difference between planned Completion and the Completion Date at a given point in time.

As such, and to avoid any ambiguity, this period ought to be clearly defined within a programme as a separate activity (typically by way of a 'hammock' or 'level of effort' bar).

As discussed in more detail in Chapter 7, planned Completion is the reference point for adjusting the Completion Date. If owing to a compensation event planned Completion moves to a later date than shown on the Accepted Programme (as adjusted), the Completion Date must be adjusted. As we shall see in Chapter 7, this has the consequence that 'terminal float' is owned by the *Contractor*, as is made clear in the NEC4 User Guide, Managing an Engineering and Construction Contract Volume 4, and Practice Note in relation to clauses 31.2 and 63.5.

As a consequence, if the *Contractor* has programmed (and is taken to have priced) to finish early, or risks it had to allow for do not materialise, it would then be inequitable for an excusable risk in the form of a compensation event to subsequently consume this time contingency. This is time that the *Contractor* has programmed for against the exposure to delay damages and would require the *Contractor* to retain its resources in place for a longer period than they would have been deployed.

The *Contractor* is entitled to finish early if they can and, as we shall see, the *Contractor* is protected from the *Client* taking advantage of that. The requirement to show planned Completion also imposes a requirement on the *Contractor* to realistically show planned Completion as it is forecast and to take into account any delays and changes effecting the date.

4.5.10 Float

Separate to the above concept is what is commonly referred to as 'float' but what more correctly is termed as 'total float'. This is a measure of the extent to which a programme activity can be delayed *without* affecting planned Completion.

Float is generated within planning software as a consequence of the manner by which the discrete activities defined have been produced and logic linked. These logic links will determine the critical path[26] through to planned Completion and also the extent to which (total) float exists in those operations elsewhere within a programme which are not on the critical path.

To avoid any ambiguity, the values for total float present at a given point in time ought to be clearly defined and displayed within a programme as a separate column. However, realistic planning is key to making proper use of the benefit of understanding the float present.

As we shall see in Chapter 7, unlike the 'terminal float', this type of float is shared and is available to accommodate either the effects of a compensation event or a lack of progress/re-sequencing by the *Contractor.*

4.5.11 Time risk allowance

The contract requires the *Contractor* to demonstrate what elements of risk that it holds has been applied to a given activity when ascertaining its duration.

Typically, when preparing a programme, risk is considered when determining the duration of activities. However, such durations are ordinarily planned on optimal conditions and so are reflective of the best that could be achieved. The NEC4 ECC by contrast is not asking the *Contractor* to give the optimal

[26] This is defined as the longest continuous path from start to finish (Completion) through a programme and hence the shortest possible time in which the project can be completed. By definition, this would result in the total float calculated being zero on the basis that any delay to the critical path would give rise to a delay to Completion.

durations but the realistic duration having considered the likely risks which the *Contractor* holds that could impact the progress of a specific activity. The *Contractor* is assumed to have priced the risks it holds to achieve Completion by the date for planned completion shown on its programme. Time risk allowances are owned by the *Contractor.*

Once again, therefore, to avoid any ambiguity, the values for time risk allowances allocated ought to be clearly defined and displayed within a programme. This could be as a separate activity or as a separate column showing what element of the overall duration is allocated to time risk allowance. The method of using a separate column to identify the time risk allowance for each activity is recommended by the NEC4 guidance to *Project Managers*.

An alternative means of incorporating time risk allowance into a programme often implemented (albeit not recognised by the NEC4 ECC) is where an allocation for time risk allowance is added as a cumulative time period at the end (known as the 'buffer'). This has the effect of driving planned Completion at a point in time in the future that would correspond to that if the time risk allowance was applied to individual activities.

This latter approach, however, lacks the sophistication or robustness of the former.

In terms of the management of risk, there would exist an element of security if the critical path activities were shown to contain a degree of time risk, demonstrating that the Completion Date is achievable even upon the occurrence of unforeseen (non-excusable) events in the progress of the works by the *Contractor.* The *Project Manager* will potentially not accept a programme that does not show realistic risk allowances.

The NEC4 User Guide, Managing an Engineering and Construction Contract Volume 4, in relation to clause 31.2, states that such allowances are owned by the *Contractor* and are retained when considering the effects of a compensation event.[27]

4.5.12 For each operation, a statement of how the *Contractor* plans to do the work identifying the principal Equipment and other resources that will be used

Equipment is a defined term and means items provided and used by the *Contractor* to 'Provide the Works'. The term 'resources' is not defined. The suggestion is clearly that the resourcing of operations should be included and an explanation of how the *Contractor* plans to do the work. Some other standard forms of contract refer to method statements.

The Guidance suggests that the *Client* may consider specifying differing levels of detail as the works progress. If the Scope does include such requirements, then the programme must be provided with that content.[28]

[27] See also Society of Construction Law (SCL) delay protocol 'Time Risk Allowance'. This is the additional time included by the Contractor within the allocated duration for an activity in a programme to allow for risks that are its responsibility under the contract. This is a contingency measure. The allowance can be zero.

[28] NEC4 User Guide, Managing an Engineering and Construction Contract Volume 4. See: https://www.neccontract.com/products/contracts/nec4/user-guides-and-flow-charts/nec4-managing-anengineering-and-construction-contract

4.5.13 Statement of how the *Contractor* plans to do the works

Historically, this related to the preparation of a supporting narrative describing in words what was shown within the programme.

However, planning software has since developed to the extent that the programme itself can demonstrate the majority (if not all) the information that the statement seeks to provide. It is possible to show resources on the programme and, in particular, planned labour levels, plant/equipment and material quantities. In doing so, it is possible to assess the demands placed on the labour, plant and materials and assess the extent to which that would be achievable.

Whether the statement is limited to identifying Equipment and resources maybe be a moot point; the wording 'a statement of how the *Contractor* plans to do the work' seems broader than this alone and could include explanations of how the assumptions made in the logic of the programme relate to the methods to be used as well as the resources.

On an ongoing basis (and as further described within Chapter 5), there may be benefit in still preparing an accompanying written narrative that sets out, at a given point in time, the basis for the remaining works, the critical path and any changes made since the last accepted programme. This will provide clear visibility to all parties as to what the programme submitted for acceptance is showing (which should in turn also facilitate the acceptance process on the basis that it allows the *Project Manager* to clearly see how a programme has developed).

4.5.14 The order and timing of the work of the *Client* and Others as last agreed with them by the *Contractor* or, if not so agreed, as stated in the Scope

In addition to programming its own work, the *Contractor* must programme work of the *Client* or *Others*.

'*work*' is not defined but is probably a reference to physical work or some service.

'Others' is a defined term.

'11.2(12) Others are people or organisations who are not the *Client*, the *Project Manager*, the *Supervisor*, the *Adjudicator* or a member of the Dispute Avoidance Board, the *Contractor* or any employee, Subcontractor or supplier of the *Contractor*.'

As mentioned above, the *Client* should specify in the Scope the base line of what will be required. We will see when we consider the compensation events in Chapter 7 that there is a relationship between the Scope and the Accepted Programme when the *Client* or Others are responsible for delay. The drafting assumes that the programme will deal with the timing of the contributions of the *Client* or Others or work. For the risk management procedures to work, the programme should make some allowance for the work of the *Client* or Others.

The compensation event clauses refer to:

60.1 (3) The *Client* does not provide something which it is to provide by the date shown on the Accepted Programme.

The *Contractor* must review the Scope and programme for anything that the *Client* is to provide.

> 60.1. (5) The *Client* or Others:
>
> - do not work within the times shown on the Accepted Programme,
> - do not work within the conditions stated in the Scope or
> - carry out work on the Site that is not stated in the Scope.

In so far as the *Client* or Others carry out work on Site that is not stated in the Scope or do not work within the conditions stated in the Scope, there is a clear risk allocation to the *Client.*

The references to the Accepted Programme are less clear cut in that we must determine if nothing is said in the Scope whether the *Contractor* is *bound* to programme for them.

It is clearly in the *Contractor's* interest to programme for these events so as to have the protection of the compensation event mechanism.

However, we come back to the question of whether the *Contractor* can unilaterally include such times within the programme without agreement if nothing is said in the Scope. If a statutory body, for example, will not agree when work will be done, which it is expected to carry out that is not on the site but impacts on completion, can the *Contractor* simply include timing in its programme?

One analysis is that it cannot and so it will be carrying the risk of both delay and additional cost if agreement cannot be reached with the statutory body. If the *Contractor* is carrying risk, then this should be covered in the time risk allowance in the programme. We will see that as a time risk allowance the delay risk is held by the *Contractor* to the extent that the delay exceeds the allowance.

Unlike the CIOB Time and Cost Management Contract TCM15, there is no drafting that permits time risk allowances to be included by the *Client* on a provisional basis such that the *Client* owns the time risk allowance and the planned completion adjusted in the Client's favour if the time risk allowance is not used.[29]

4.5.15 The Dates when the *Contractor* will need access, acceptances, Plant/Materials provided by the *Client*, information from Others

> The conditions also refer to:
>
> - the dates when, in order to Provide the Works in accordance with the programme, the *Contractor* will need
>   - access to a part of the Site if later than its *access date*,
>   - acceptances,
>   - Plant and Materials and other things to be provided by the *Client* and
>   - information from Others.

[29] Clause 45.3.

The *access dates* are stated in the Contract Data part One. The dates are relevant to clause 30.1, which states that the *Contractor* does not start work on Site until the first *access date*. The *Contractor* can programme to start work after the *access dates*. The *Client's* obligation regarding access is dealt with in clause 33.1 and also through the compensation clause 60.1(2), which provides that it is a compensation event if 'The *Client* does not allow access to and use of each part of the Site by the later of its *access date* and the date for access shown on the Accepted Programme.'

The word 'acceptances' is not defined but is used in the context of communications with the *Project Manager* who is responsible for accepting or not accepting submissions. For example, the *Contractor's* design must be submitted for acceptance. This is distinguished from 'approvals', which is not defined but could be required by the Scope from Others or the *Client*.

Plant and Materials are defined terms and are items intended to be included in the permanent works. Unless the Scope has stated that Plant and Materials are to be provided by the *Client* then the *Contractor* would be responsible for procuring them.

'Information' is not a defined term and we have seen 'Others' is defined.

This broad category 'as described above' relates to matters that the *Contractor* is reliant upon, outside of its control, in order to work to the programme. Such provisions will, therefore, relate to any *Client* or third-party interfaces (apart from work), which would have the potential to affect the *Contractor's* works.

Typically, these are shown by way of either an activity or a milestone to demonstrate and highlight the date that the *Contractor* needs an action from the *Project Manager* or something the *Client* or Others provide (be it the acceptance of a design by the *Project Manager* or the provision of a free issue material by the *Client*.

Any periods included ought to be in accordance with the periods set out within the Scope where defined. In addition, an adequate level of detail (with defined elements and logic links to the subsequent works impacted) is required in relation to these items in order to be able to model any potential delays attributed to these dates or periods not being achieved.

This will result in an integrated programme being available for the benefit of all with the interfaces clearly demarked and identifiable at any given point in time. This will, in turn, assist the parties in completing the programme and works on time.

4.6. Other commercial considerations

As outlined above, the way in which the *Client* has drafted the Scope will impact upon the *Contractor's* planning and the risks that it must allow for. Failure to capture risk in the programme may well leave the *Contractor* holding the risk under the contract.

4.7. The relevance of the programme to compensation events

As we have seen, the Accepted Programme is relevant to the allocation of risk that equally benefits the *Contractor*. If we consider the drafting related to the compensation event mechanism, a number of clauses cross refer back to the Accepted Programme.

Those clauses that make direct reference to the Accepted Programme are set out below.

> 60.1(2) The *Client* does not allow access to and use of each part of the Site by the later of its *access date* and the date for access shown on the Accepted Programme.
>
> 60.1(3) The *Client* does not provide something which it is to provide by the date shown on the Accepted Programme.
>
> 60. 1(5) The *Client* or Others do not work within the times shown on the Accepted Programme.
>
> 60.1 (19) An event which:
>
> - stops the *Contractor* completing the whole of the works or
> - stops the *Contractor* completing the whole of the works by the date for planned Completion shown on the Accepted Programme, and which
> - neither Party could prevent,
> - an experienced contractor would have judged at the Contract Date to have such a small chance of occurring that it would have been unreasonable to have allowed for it and
> - is not one of the other compensation events stated in the contract.

If the Accepted Programme is lost, so are these references points. See Chapter 11. Beyond simply addressing all the bullet point items at clause 32.1 (as set out previously), the programme must also meet other requirements if it is to be accepted by the *Project Manager*. This process is set out in more detail in Chapters 5 and 6, but the broader requirement is to produce programmes for submission that meet the tests defined at clause 31.3 (as set out below).

> 31.3 Within two weeks of the *Contractor* submitting a programme for acceptance, the *Project Manager* notifies the *Contractor* of the acceptance of the programme or the reasons for not accepting it. A reason for not accepting a programme is that
>
> - the *Contractor's* plans which it shows are not practicable,
> - it does not show the information which the contract requires
> - *it does not represent the Contractor's plans realistically or*
> - *it does not comply with the Scope.*

The programme in addition to complying with the letter of the contract (bullet point 2) and the Scope (bullet point 4) must be practicable and realistic. Such terms are everyday terminology.

The *Contractor's* plans that it shows are not practicable refer to specific content that is shown and is not practicable. Note the reference to the *Contractor's* plans. These plans must comply with any contractual dates for access or other constraints on delivery imposed by the Scope. For example, the dates the *Client* might have identified for design submissions or acceptances. But if the Client cannot then comply with such dates, that would not be a reason for rejecting the Programme as not practicable.

The wording "it does not represent the *Contractor's* plans realistically" may refer more generally to an appraisal of the programme including assumptions – for example, risk allowances or durations could be rejected as unrealistic.

The interpretation of such provisions may be open to some debate[30] but the intention behind the drafting of the contract is that the programme is intended to be an up-to-date view of the actual progress of work in the context of the constraints that are then known, with a realistic position regarding planned completion.

If this central purpose is kept in mind, then it should be possible for the acceptance of the programme to be matter of everyday management. It requires, however, that there is trust in the programming and that the *Contractor* is transparent in its approach to preparing the programme and when updating its programme to reflect the progress of the project properly and fully.[31]

In line with this, the programme should not hide delay or fail to show risk allowances. Nor should it seek to unrealistically represent the relationship between operations.

We will come back to this in the sections on revising and updating the programme.

4.8. Additional obligations arising from the Z clauses

The conditions of contract envisage that additional requirements for the programme are dealt with in the Scope. The *Client* may, however, wish to change drafting in relation to the process, such as the relationship between the acceptance of the programme and payment or the acceptance procedures more generally, in which case Z clauses would need to be used. Any such drafting should be done with care and with the full purpose of the programme kept in mind.

The drafting of clause 50.5 permitting the withholding money in limited circumstances where a first programme containing the information the contract requires has not been submitted is of limited use. It does not apply to the revision of programmes nor in the case where a first programme was accepted as part of the *Contractor's* tender.

Other forms of contract use devices such as recovery of costs incurred by the client in programming or making the right to carry out work subject to the acceptance of the programme. While contractual mechanisms have their place, the best incentive for the parties to provide the programming input the contract requires is to understand the benefits and downsides of providing or not providing a programme.

Of equal importance is to recognise the considerable resource that is required to programme as required and to have a clear-eyed appreciation of the detail required to meet the contractual requirements.

4.9. Tender evaluation

At the point where the *Client* is preparing the tender documents, there is an opportunity both in terms of the approach to the tender process and the contractual requirements to approach these issues in such a way that the programming will be able to fulfil its functions under the contract.

---

[30] See *Keating on NEC Contracts, 2nd Edition* (Thomas and Lee, 2022: p. 45) and Eggleston (2006: p. 157) *NEC3 Construction and Engineering Contract, 2nd Edition.*

[31] Note that the FIDIC and CIOB forms of contract contain extensive provisions dealing with the provision of progress records to ensure that revised programmes are based upon actual progress.

As seen in the guidance document Volume 2, the contract highlights specific areas that the *Client* should consider when drafting the Scope. It is also important to remember that by accepting a programme at the tender stage certain obligations may become fixed by reference to that programme.

There is no obligation imposed on the *Client* or the *Contractor* to carry out the work strictly in accordance with the programme, but the *Contractor* is expected to identify specific matters from the Scope or by agreement with the *Client* or *Others* that are relevant to achieving the planned completion date.

For this reason, it is suggested that the programme review at the tender stage is undertaken on the same basis that it will be undertaken by the *Project Manager* post contract.

A tender that contains a programme that does not meet the contract test is potentially storing up additional cost of resourcing post contract and imposing impracticable requirements on the *Client* as well *as* undermining the procedures for risk management in the contract.

In addition to the programme (if submitted at the tender stage meeting the contract test), the tender evaluation may allow broader considerations and scoring of the programming ability of the contractors. A contractor whose programme has fully understood the requirements of the NEC contacts for programming, and who has in place the proper planning resource and management systems to update programmes and report progress in an accurate and transparent way, is fundamentally better placed to deliver a project in a timely way and to give the greatest potential for minimising the impacts of risk than one who has not.

Leaving such matters to be addressed and developed post contract by the *Project Manager* is not a good starting point for the parties and places a considerable administrative burden on the *Project Manager.*

A client who considers that a programme that meets the contract requirements is unnecessary or of little value should consider whether the NEC ECC short form of contract is more appropriate or amend the contract.

4.10. Pricing

The NEC family provides for several pricing options. Where the option chosen is Option A, which requires an activity schedule (the Activity Schedule) to be produced by the *Contractor,* the *Contractor* must consider the relationship between the pricing document and the programme. The important point to note is that in the option A form of contract there is a relationship between the programme and the Activity Schedule. See clause 31.4, which requires the *Contractor* to provide information that shows how each activity relates each operation on the programme.

The *Contractor* is paid based on completed activities. Each activity is given a price. If the Activity Schedule does not expressly include an activity it is deemed to have been allowed for in the prices of other activities. The contract envisages that the *Contractor* is pricing all the content in the Scope and the risks and constraints and is also programming to complete that content.

The Option A contract conditions envisage that there is some reasonable alignment of the Activity Schedule and the programme and seeks to maintain that position. If the Scope changes or the Contractor changes its method of working, then the programme changes and the Activity Schedule must change.

Operations on the programme and activities in the Activity Schedule should be capable of clear cross referencing.

Where the Scope changes this will usually be a compensation event and the Activity Schedule is adjusted following the assessment of the compensation event.

But there are other circumstances where the Activity Schedule must be altered.

> 55.1 Information in the Activity Schedule is not Scope or Site Information. If the activities on the Activity Schedule do not relate to the Scope, the *Contractor* corrects the Activity Schedule.
>
> 55.3 If the *Contractor*
>
> - changes a planned method of working at its discretion so that the activities on the Activity Schedule do not relate to the operations on the Accepted Programme or
> - corrects the Activity Schedule so that the activities on the Activity Schedule relate to the Scope
>
> the *Contractor* submits a revision of the Activity Schedule to the *Project Manager* for acceptance.

Clause 55.1 states that the Scope identifies the work content, and the Site Information identifies information relevant to pricing the physical conditions. The Activity Schedule cannot qualify these documents. If the Activity Schedule does not reflect that content, it must be corrected. The correction does not allow the *Contractor* to correct its mistake in pricing.

Clause 55.4 prevents adjustment of the Activity Schedule that would seek to circumvent such problems or is not related to the operations in the programme. The *Project Manager* would not accept the proposed change to the Activity Schedule if

- it does not relate to the operations on the Accepted Programme,
- any changed Prices are not reasonably distributed between the activities which are not completed or
- the total of the Prices is changed.

A contractor must be aware that in pricing risk and programming for it there should be a consistent approach. An activity schedule that is based upon a realistic programme is important to avoid problems with payment. It should be possible to relate the work breakdown structure to the Scope and the Activity Schedule.

Willis T
ISBN 978-1-83662-481-3
https://doi.org/10.1108/978-1-83662-478-320251006

Chapter 5
Updating the programme

5.1. Introduction

The previous chapter described the requirements in relation to preparing a programme under the NEC4 ECC form of contract. In addition to these provisions NEC4, ECC also sets out a number of requirements in relation to the management of the programme on the basis that the contract necessitates that it is to be reissued for acceptance throughout the duration of the project.

This chapter will consider in more detail how the programme is to be revised.

5.2. The (first) Accepted Programme

Before reviewing the criteria set out in NEC4 ECC in relation to updating a programme, it is useful to first review clause 31.3 and the process by which an Accepted Programme comes into existence (after which then it is to be updated). The basis for producing a programme capable of being accepted is predominantly set out under clause 31.2 (and has been described within the previous chapter).

As set out by way of clause 31.3, the *Project Manager* is required to determine if a programme by the *Contractor* is capable of acceptance within 2 weeks of its submission. If they fail to respond within this time period, either communicating acceptance or non-acceptance, NEC 4 ECC has introduced a provision that the *Contractor* is able to notify the *Project Manager* of such failure.[32] If the failure continues for a *further* week, then the submitted programme is treated as having been accepted (clause 31.3).[33]

If a programme is not accepted, the *Project Manager* is required to provide its reasons for not doing so (with the intent then that the *Contactor* could review and address these comments in future submissions).

Clause 31.3 sets out four reasons that entitle the *Project Manager* to not accept the programme:

1. **The *Contractor's* plans are not practicable:** this could include what is deemed to be incorrect sequencing, overlapping of works giving rise to practical impracticalities or insufficient activity durations.
2. **It does not show the information which this contract requires:** this could include criteria such as float, time risk allowances, the relevant access/Key Dates, or the provision of information/materials by others.
3. **It does not represent the *Contractor's* plans realistically:** this could include the *Contractor* failing to properly identify the scope of the remaining work, inadequate/contradictory resources

[32] Under NEC3, the *Contractor*, following a failure by the *Project Manager* to respond, was only able to notify a compensation event for a lack of response. The *Contractor* still can notify of such failure in accordance with clause 60.1(6).

[33] Based on this wording, it would follow that this newly introduced sanction within NEC4 ECC will only apply if the *Contractor* operates the clause, and the *Project Manager* still fails to respond.

shown to be undertaking the remaining work, or the sequencing for the remaining work not matching that which is being undertaken.

4. **It does not comply with the Scope:** the programme is considered to not reflect the full scope of work scope for the project or other requirements stated in the Scope.

If the *Project Manager* chooses not to accept for any other reason than those set out above in the contract then, by default, that would give rise to the occurrence of a compensation event as defined by clause 60.1(9).

5.3. When is the programme to be updated?

The contract has a specific section dealing with the revision to the programme.

Clause 32.1 states what must be shown on each revised programme, and it is this which is central to the process and must be considered first.

5.4. Requirements in relation to the updated programme

Clause 32.1 (Revising the Programme) sets out the requirements in relation to submitting updates to the Accepted Programme. These set out further obligations upon the part of the *Contractor* in relation to the level of planning required.

Factors relevant to the revision of the programme are set out at clause 32.1 and here it is stated that the *Contractor* is required to show on each revised programme the following:

- The actual progress achieved on each operation and its effect upon the timing of the remaining work,
- How the *Contractor* plans to deal with any delays and to correct notified Defects,
- Any other changes which the *Contractor* proposes to make to the Accepted Programme.

Further consideration has been given to each of these elements defined at clause 32.1, in turn, below.

5.4.1 Actual progress achieved

A revised programme is required to be fully updated with the progress achieved at the cut-off date for the programme (which itself would align with the requirements set out at clause 32.2) for each and every activity. This will, in turn, enable (i) any delays to be identified, (ii) any of sequence work or change in resources or methods of working to be determined, and (iii) the remaining work to be properly identified.

As such, the revised programme will set out the *Contractor's* delays together with a contemporaneous assessment of the impact of these on the remaining work and, as we shall see, the *Contractor's* plans (see below).

A failure to show this would give the *Project Manager* grounds to reject the programme under clause 31.3 as not showing information required by the contract.

The consequences of the programme not being accepted are discussed more fully in Chapter 11 but will have a significant impact on risk allocation and the potential for disputes.

5.4.2 How the *Contractor* plans to deal with any delays and correct notified Defects

As we saw above, the actual progress achieved will show any delay that has impacted at the time of the programme but the drafting here and the use of the word plans relates to the future risks of delays owing to risks that the *Contractor* holds.[34] The correction of notified Defects must be shown as it is an operation that has to be allowed for as it is a *Contractor's* risk.

There has been a change to this clause from that previously set out within NEC3 ECC. This previously expressly provided for the inclusion of the effects of implemented compensation events matters as a separate requirement.[35] These do not need to be shown under NEC4. We shall see that the compensation event process provides the record of the effect of implemented compensation events on the Completion Date and the Prices. In that process, the programme will either have been revised on an agreed basis or there will be no Accepted Programme.[36]

The removal of this requirement has reduced a potential source for confusion, as will be shown in Chapter 7.

5.4.3 Any other changes the *Contractor* proposes to make to the Accepted Programme

Whilst the first two bullets relating to progress and delays relate, on the whole, to events that have taken place, this specific requirement has a greater potential to relate to matters that impact upon the activities relating to the remaining work. Similarly, matters here may also be a reflection of the change necessary to the programme to reflect the impact of progress and delays on the plans of the *Contractor*.

For example, if delay in an activity has the consequence that work in part of the site will not now commence on the *access date* as shown on the existing programme, the access date shown on the programme is required to be changed. If the *Contractor* changes its method of working or resourcing assumptions these would also need to be shown. If a risk allowance has not been used and an activity has completed, this would also potentially change the Accepted Programme.

Where the programme changes, this does not solely effect the dates on which the *Contractor* plans to carry out operations. It can also impact on dates on which the *Client* has to provide things. In addition, if the *Client* does not meet these requirements, it can give the *Contractor* an entitlement under the compensation event process. There are several specific compensation events that cross reference the Accepted Programme. These include:

- Clause 60.1(2): The Client does not allow access by the later of the *access* and the date for access shown in the Accepted Programme.
- Clause 60.1(3): The *Client* does not provide something by the date shown in the Accepted Programme; and

[34] Risks held by the *Client* are not programmed for until they actually impact progress and are captured under the first bullet point. Any future *Client* risk if it materialises will be a compensation event.

[35] NEC 3 32.1: The *Contractor* shows on each revised programme. the actual progress achieved on each operation and its effect upon the timing of the remaining work, the effects of implemented compensation events, how the *Contractor* plans to deal with any delays and to correct notified Defects and any other changes that the *Contractor* proposes to make to the Accepted Programme.

[36] See Chapter 7 and Chapter 10.

- Clause 60.1 (5): The *Client* or others do not work within the times shown on the Accepted Programme.

5.4.4 Limitations on the changes the *Contractor* may make

The *Contractor* is not permitted to unilaterally re-programme works without regard to the requirements for acceptance and the constraints that the contract imposes.

For example, the *Client's* obligation with regard to giving access is to give access by the later of its *access date* and the date for access shown on the Accepted Programme.

A *Contractor* cannot re-programme and require the *Client* to provide access by an earlier date than previously shown on the Accepted Programme. The *Contractor* must also comply with the constraints identified in the Scope.

The process for acceptance in accordance with clause 31.3 requires that any re-programming undertaken meets the requirements that the *Contractor's* plans are both 'practical' and 'realistic'. The Project Manager must assess (prior to accepting) that any revised dates for which the *Contractor* states it requires access, acceptances or plant/materials/information to be provided by the *Client* (or Others) are realistic and practicable and in line with the Scope. A *Client* should not be required to provide anything earlier than it had agreed to nor can the work of the *Client* or Others be unilaterally programmed to earlier dates. Where the *Client* or Others cannot meet dates then new dates should be agreed and instructed as compensation events. The *Project Manager* must be alert to such risks and give early warnings to mitigate them. If revised dates are accepted but not then met, a further compensation event would arise.

5.4.5 Submissions of a Revised Programme

Clause 32.2 of NEC4 ECC sets out when the *Contractor* is to submit a revised programme.

> The *Contractor* submits a revised programme to the *Project Manager* for acceptance.
>
> - within the period for reply after the *Project Manager* has instructed the *Contractor* to,
> - when the *Contractor* chooses to and, in any case,
> - at no longer interval than the interval stated in the Contract Data from the starting date until Completion of the whole of the works

The requirement to update the programme has to be considered with what it must show. The starting point is that the *Contractor* should update the programme where the Accepted Programme does not represent what is now the position. The *Contractor* should submit a revised programme when the facts have changed and so this is not given a timescale but refers to 'when the *Contractor* chooses'. If the *Project Manager* considers the programme is not now up to date then the *Project Manager* can instruct the *Contractor* to do so and lastly and in any case, in accordance with the intervals stated within Part 1 (Section 3) of the Contract Data. This is set by the *Client* in Contract Data Part 1. The interval should not be too great.

This provides a default position where if nothing changed there would still be an Accepted Programme that is up to date and records that nothing has changed and gives the agreed position with regard to progress and planned completion.

> There is only one Accepted Programme at any point in time. The definition makes this clear:
>
> 11.2 (1) The Accepted Programme is the programme identified in the Contract Data or is the latest programme accepted by the Project Manager.

But the various superseded 'Accepted Programmes' must be archived and form a progress record for the project. They will also be referred to in order to assess compensation events.

5.4.6 The consequences of not providing a revised programme

We will consider in much more detail in later chapters the implications of not having an Accepted Programme but it has unwelcome consequences for the Parties and increases the burden on the *Project Manager*. Failure to revise the programme as required will mean that there is then no Accepted Programme to use for the compensation event process (see Chapter 11).

A simple reading of the contract clause 10.1 states that if the *Contractor* fails to submit a programme in accordance with the contract, then, on the basis of clause 10.1 and the requirements set out at clauses 31 and 32, the *Contractor* is in breach of contract and may also be subject to money being withheld where clause 50.5 applies.

Notwithstanding the detailed provisions and obligations set out within the contract in relation to programme, unless the *Client* wishes to terminate the contract in light of this breach,[37] there is no other obvious practical remedy available. A claim in damages may be of little benefit to the *Client* and it is unlikely that the *Contractor* could be forced to provide a programme. Having an Accepted Programme benefits both Parties and a competent *Contractor* should not require forcing to produce and update the programme.

Clause 64.2, which relates to the assessment of a compensation event by the *Project Manager* in the absence of an Accepted Programme (or for other reasons) to model the effect of a delay to substantiate the adjustment to the Completion Date, provides implicitly for the preparation of a programme by the *Project Manager* to reflect the remaining work. We consider in more detail later in Chapter 11 how the *Project Manager* might approach that task but any such programme is not capable of being adopted as the Accepted Programme.

5.4.7 Purpose of updating the programme

There are many reasons why the programme needs to be updated and proper resourcing should be made available to allow this to happen.

5.4.7.1 Effective planning and collaboration

The contract requirements relating to the revision of the programme are stated at clause 32 and the intention is that the *Contractor*, in order to fulfil the requirements relating to revising the (Accepted) programme, must show actual progress and changes to its programme in addition to meeting the general requirements of the programme set out at clause 32.1). This requires transparency from the *Contractor*. As we shall see, the *Contractor* benefits as does the *Client*.

[37] See clause 91.2; whether this would meet the required test is debateable and is in any event a very extreme remedy.

The rationale for doing so is the intention that the *Contractor's* programme is envisaged to be used as a proactive tool for both managing the works on a day-to-day basis, but also as a mechanism to demonstrate and forecast the effects of change and the *Contractor's* entitlement to an adjustment to the Completion Date.

Where the Accepted Programme has become out of date on the basis that revised programmes have not been submitted or instructed or if when submitted the programme have not been accepted, there exists a risk to the parties that the works are not properly planned, the parties' requirements are not communicated and there is no basis to accurately forecast when Completion might be achieved and whether the Completion Date is achievable.

As we shall see in later chapters, for both parties, uncertainty increases and the possibility of a complex dispute becomes more likely. Accordingly, NEC4 ECC provides a set of programme management requirements that, if followed, will benefit all the parties.

5.4.7.2 Compensation events

The role of the programme in assessing delay for the purpose of compensation events is set out in detail in Chapter 7.

As noted above, NEC4 ECC has removed from the drafting in the ECC the express requirement to show the effect of any implemented compensation events within a revised programme. The effects of implemented compensation events are captured in the adjustment to the Prices and Completions Date through the process described in detail in Chapter 7.

5.4.7.3 Early warning notifications

The early warning procedure places a requirement on both parties to notify the other if they become aware of any matter that could affect time, cost or quality (in accordance with clause 15.1 of the NEC4 ECC).

The process relates to issues that *may* among other consequences have an impact to delay Completion. The purpose is to encourage parties to raise early warnings where there is an issue envisaged. The parties then can discuss the issue and collectively seek to minimise (or eliminate) the potential effects.

On the basis that this form of contract places risk management at the heart of the contract and the programme is central to the decision-making process when responding to risk, there is a real benefit that the Accepted Programme provides a basis to discuss the potential effects of matters raised in the early warning notifications. The programme is not itself a 'notice' (see clause 13.7) and there is a detailed requirement to maintain an Early Warning Register to monitor *potential* issues that could increase the Prices or delay the Completion Date impair the quality of the works in use.[38]

Under the NEC3 ECC, part of the early warning process required a 'Risk Register' to be produced post contract. The use of the term Risk Register in the NEC contracts caused a good deal of confusion and in NEC4 it is now referred to as the Early Warning Register. The document does not provide the basis for

[38] See clause 15.1.

any action to be taken under the contract. It simply records decisions post contract in response to risks and the decisions must then be actioned either by agreement or instruction. The Early Warning Register does not allocate risk or assign probability of risk or its consequences.

The contract provides for risk allocation in the compensation event clauses and the *Client* liabilities.

A contractor must allow for risk that it holds in its programming, but it is not required to programme for risks that are held by the *Client*.[39] Since Early Warning Notices may be raised for matters that could be either party's risk in the contract, the Early Warning Notice itself does not determine whether the programme should take account of the risk. If the risk in question is held by the *Contractor* then the programme should take it into account in the operations potentially impacted by it. The contract also distinguishes between the trigger for an early warning and the trigger for risk more generally.

Clause 15.1 provides:

> The *Contractor* and the *Project Manager* give an early warning by notifying the other as soon as either becomes aware of any matter which could
>
> - increase the total of the Prices,
> - delay Completion,
> - delay meeting a Key Date or
> - impair the performance of the works in use

The word 'could' is not the same as risk that 'has a significant chance of occurring', which is the terminology the contract uses for risk post contract. Or the phrase used in the context of some compensation events: 'risk that has such a small chance of occurring at the Contract tender stage that it would have been unreasonable for an experienced contractor to have allowed for'.

Anecdotally, it is suggested that under the NEC3 contract some *Contractors* refer to Early Warning Notices in their programme, but this is not a contract requirement and is confusing and unnecessary as the Early Warning Register provides the record of what risks have been the subject of the early warning notices, and the actions agreed in the required Early Warning meetings are also recorded in the Early Warning Register.

There are some implications for the programme in that if the *Contractor* holds the risk and seeks to mitigate its impact then that should be shown in a revised programme as the new sequence of work or resources that it now intends to employ and the effect it will have on *planned completion*.

[39] If the *Client* has included in the Scope a period when work is to be done by others the *Contractor* will allow for that work, but it does not hold the risk of that work taking more time than allowed. There is no mechanism to allocate time risk allowances to the *Client* unlike some other standard forms. See CIOB Time and Cost Management Contract Conditions TCM15 (2015, clause 45).

If the *Client* holds the risk, then any intervention[40] – going beyond simply allowing the risk to materialise – would have to be made be through an instruction or by some documented agreement, in which case the *Contractor* would also have to revise the programme to show the new sequence of work or resources that it now intends to employ and the effect on planned completion. The risk is then captured in the compensation event mechanism and any delay owing to it would be shown in the programme as discussed in more detail in Chapter 7. If there is no intervention but the event starts to impact the programme or the remaining work then it would also have to be taken account of in the programme. The *Contractor* must then use the compensation event mechansim to adjust the Completion Date.

5.4.7.4 Acceleration

The acceleration provisions set out at clause 36 of the NEC4 ECC are a means of bringing forward (in time) the Completion Date or Key Dates by agreement.

Where the Completion Date might not be met owing to delay on the part of the *Contractor,* as we have seen, there is no right to force the *Contractor* to meet a specific level of progress or supplement its resources. Similarly, when delay is caused by a compensation event, the *Contractor* has no obligation to prevent the delay or reduce its impact. The early warning procedure in the circumstance of a risk that the *Client* holds gives the *Client* the opportunity to instruct a change in response, but provided the *Contractor* gave an early warning its obligation is to respond to a compensation event in which response is based on specific contract assumptions. These are within NEC4 ECC that the *Contractor*, when addressing a compensation event, is assumed to[41] respond to competently and promptly and that the resources used in response are reasonably incurred. The *Contractor* must also allow for the risk of events that have a significant chance of occurring and are not compensation events.[42] This is in keeping with general legal principles. The *Contractor* is only therefore required to employ measures that do not give rise to additional cost to it beyond the requirements stated. If, having applied those assumptions, the planned completion date becomes a later date than that shown on the relevant programme then that is the date by which the Completion Date must be adjusted. There is no obligation to achieve any earlier date.

Acceleration is where an earlier date is desired but the wording of clause 36.1 is clear that it is left to the Parties to consider and mutually determine if they wish to engage in the notion of the implementation of acceleration measures.

The *Contractor* is under no obligation to engage in the process if it does not wish to. If the *Project Manager* requests a quotation for acceleration, the *Contractor* is not required to provide a quotation. If it does provide a quotation for acceleration, the *Project Manager* must either accept the quotation for the agreed acceleration period and hence change the Completion Date and (if necessary) Key Dates or not accept, in which case the Completion Date and Key Dates will remain as it was on the last Accepted Programme.

Should the quotation for the agreed acceleration period be accepted then it follows that this would be incorporated into the revised programme to be submitted for acceptance.

---

[40] Any such intervention would be assessed as a compensation event under 60.1(1) or 60.1(4) unless the Parties had agreed a response to the risk under clause 12.3.

[41] See clause 63.8.

[42] See clause 63.9.

Willis T
ISBN 978-1-83662-481-3
https://doi.org/10.1108/978-1-83662-478-320251007

Chapter 6
The role of the *Project Manager*

6.1. Introduction

The *Project Manager* is identified in the Contract Data Part One or any replacement person notified to the *Contractor*. The person is appointed by the *Client* and carries out the procedures that are part of the contract administration. The *Project Manager* is also the person responsible for any instructions or interventions on the *Client's* behalf, including the response to risk. The Accepted Programme is intended to be a central tool in that process. In this chapter, we will consider how the *Project Manager* plays a central role in the programming by accepting or not accepting programmes.

6.2. The role of the *Project Manager* general considerations

As we have seen in the preceding discussion, the contract drafting seeks to keep in place an Accepted Programme. The purpose of the programme is to provide a baseline in terms of planning the actions of the *Contractor* the *Client* and Others and to permit the potential impact of risk to be better understood and analysed.

There is no direct link between the Early Warning procedure and the programme, but it is implicit that the Accepted Programme should have a vital role in understanding the implications of risk and the implications of potential mitigation or avoidance of risk.

The Accepted Programme should also provide an agreed record and factual basis for the decisions being reached. It limits the scope for future disagreement. As we have seen, the NEC4 ECC makes no other specific provision for progress meetings or for progress records.

If the Accepted Programme is lost, then the *Project Manager* will have to resort to other mechanisms to establish progress and to consider the impact of risk.

There is resource cost for the contract in that both parties are likely to duplicate records, information and analysis to a greater degree than would otherwise be the case.

The number of hours that the *Project Manager* would have to commit to a project are likely to be significantly increased as will be the time that the *Client* will be engaged.

Behind the bare fact that there is no Accepted Programme is the spectre of a serious potential dispute.

The programme cannot, therefore, be considered as an isolated document, but its acceptance or non-acceptance goes to the heart of the contract administration. The role of the *Project Manager* as a person appointed by the *Client* has specific obligations under the contract and more broadly in terms of competently discharging its obligations under their own appointment document. Given the centrality of the programme, in accepting or selecting the role of the *Project Manager*, serious consideration must be given to planning capability or access to it. Appointing a *Project Manager* who does not have the competence or the resource to be able to manage the process is a serious failure on the part of the *Client* and the person accepting the appointment.

The demands of the role can be better managed by careful thought at the tender stage as to who will carry out the function and whether additional content should be included in the Scope such that both the *Contractor* and the *Client* are clear about what will be required on a specific project.

Specific requirements in relation to the programme might be:

- progress records
- software to be used
- format of presentation of information and access to native content
- identification of a dedicated contract planner as a Key Person
- any outstanding information at the tender stage from the *Client* re *Client* activities or Others and programming assumptions to be made
- narrative of programme
- explanation of critical path – time risk allowances – float
- method statements
- resource assumptions.

6.3. Specific requirements of the role of *Project Manager*

As we have seen in the previous discussion, it is the *Project Manager* who must accept or reject the programmes submitted for acceptance. The timescale is 2 weeks to accept or reject. That can be extended by agreement. It is clear, however, from what has gone before that the contractual timescales should be treated as a maximum. Timescales can be shortened by open access and a collaborative approach to programming.

The reasons for accepting or not accepting the programme are set out in the contract. If the *Project Manager* departs from those reasons that is a compensation event. The *Project Manager's* actions or inactions are also capable of review and correction by the *Adjudicator*.

In any such challenge, the *Project Manager* will not only have to be able to show the reason for which a programme was not accepted but also substantiate it.

Where a programme is not accepted we will also see that the programming will default to the *Project Manager*.

6.4. Acceptance of revised programmes

The *Project Manager's* acceptance is significant under the contract, yet under clause 14.1 it is essential to know that there is some protection. Clause 14.1 states that acceptance of any communication does not transfer liability away from the *Contractor* to comply with its obligations under the contract. If the *Contractor* deletes an element of work from the programme that is stated in the Scope or ignores a constraint, the subsequent acceptance of the programme does not by default overwrite the original Scope. The only person who can give an instruction to change the Scope on a project is the *Project*

Manager. Having said that, acceptance under the contract is important, as it is recognition that from the *Project Manager's* perspective that the programme is realistic, practicable and shows the information that the contract requires. The *Project Manager* cannot correct the programme but could instruct that a revised programme is submitted for acceptance.

The contract procedure for Acceptance permits 2 weeks for the *Project Manager* to review the programme from the date of its submission. The *Contractor* has the *period of Reply* in which to submit a revised programme when instructed to. These timescales can cause significant lag in keeping the Accepted Programme up to date. A short *period of Reply* for the programme should be allowed in the Contract Data.

The maximum timescales for response within the contract do not practically work on projects where there is significant change. The standard timescales within the contract must be sensible as they cover the whole of the construction industry. They cannot be too short as this will not work in numerous circumstances, and it is not desirable to have to change them with Z clause amendments for each contract. That is not necessary if on a practical level most timescales are a maximum and looked to be shortened by project teams for the benefit of all concerned. The programme acceptance period for response within 2 weeks is no exception to this. Otherwise, what can tend to happen is that a *Project Manager* will take the full 2 weeks before responding to a *Contractor* whether a programme is accepted or not.

During that 2-week period from when the *Contractor* submitted his update there is often a lot of change, and the danger is that the *Project Manager* feels that they subsequently cannot accept the programme. The *Project Manager* should be able to accept this programme as the 'line in the sand' at the time of issue, reflecting the true representation of completed and programmed works at that point. Any subsequent change to any dates will be picked up and detailed in the next programme.

However, very quickly it is possible to get out of sync in getting programmes accepted, and it becomes increasingly difficult to get back on track. This is a big problem to both parties as the Accepted Programme is the reference point for all compensation events and change that occurs. If there is not a recent Accepted Programme in place then commercially the Parties are liable to be moving further apart in terms of commercial agreement as a result of these changes.

Historically, under other forms of contract the *Project Manager* has not had to accept any revisions of the programme during the life of the contract. However, it is very important that the *Project Manager* understands that under any NEC contract they have an obligation to accept or not accept the programme (and have a reason for not accepting under the contract, otherwise it will be a compensation event). The Accepted Programme should be considered just as important to the *Project Manager* and the *Client*, as this will ensure that all Parties understand the liability of for delays as they occur in relation to the planned Completion and Completion Date milestones. The regular Accepted Programme will become a live real management tool reflecting all the activities at any one point in time.

6.5. Collaborative programming

One solution for *Clients* and project teams to assist both the *Project Manager* and the *Contractor* is to adopt in the Scope a process of collaborative programming where the programme is a shared open resource that is informally continually under review and updated through the Acceptance procedure at very quick notice once it becomes clear that the programme has changed. *Project Managers* can, in any event, seek to work collaboratively with the *Contractor* even if the *Client* has not outlined such a process in the Scope.

Where open collaborative working is envisaged, it is challenging to expect that the *Project Manager* will instigate this from a standing start post contract. The programme is just one of the many time-bound tasks that a Project Manager must address. See the Project Startup checklist in Appendix 1 to Volume 4 of the NEC4 ECC User Guide.

The question of progress reporting is not expressly dealt with in the NCE4 ECC but it will not be possible to update a programme on a continual basis without an agreed structure for progress reporting. The structure can be designed to link records clearly to the programme. The FIDIC forms of contract provide express drafting on progress reporting[43] and link the submission of progress reports to payment.

The most detailed approach to progress reporting and the programme in a standard form is found in the CIOB Time and Cost Management Conditions 2015 TCM15, clause 39 and Appendix E. This sets out a standard for linking progress reports to the 'Working Schedule' (programme). The contract envisages a common data environment using the specified software identified by the Employer.

[43] See clauses 4.20 and 6.10.

NEC4: Effective Preparation and Management of the Construction Programme

Willis T
ISBN 978-1-83662-481-3
https://doi.org/10.1108/978-1-83662-478-320251008

Chapter 7
Adjusting the Completion Date

7.1. Introduction

As we have seen, the Accepted Programme is central to risk management. The ethos of the NEC is not to adjust retrospectively to risk but to manage it. To forecast its potential impact and to reduce or avoid it if we can. To the extent we cannot prevent risk materialising, the contract is seeking to achieve an agreed basis for adjusting both our actions and contract parameters in time and cost in response to it. We will look at how the NEC4 ECC adjusts the Completion Date, but before doing so it is necessary to consider that issue more broadly.

7.2. Adjusting the Completion Date – the context in which the NEC4 ECC approaches this task

Adjusting the completion date of a contract is a central aspect of the administration of contracts. This is necessary to preserve the point at which liability arises for delay and is linked to the ability of a client to recover damages for the loss it suffers as a result of delay in completing the works. The focus in this setting is the liability in contract and since there are several principles of law that will impact upon the effectiveness of rights of recovery, the drafting of this aspect of the contract tends in most contracts to concentrate on the circumstances in which adjustment is made, rather than the detailed mechanism for the adjustment. Preserving the rights of the *Client* to potential damages does not of itself deal with the impact of delay and so we see the NEC4 ECC is also seeking to preserve the programme as an agreed basis for carrying out the work. It concentrates on process as much as liability.

Most contracts, as we have seen, do not concentrate on the process.

In the absence of detailed rules in the contract, courts in different jurisdictions have provided some guiding rules. There is considerable variety in approach. Much depends upon the interaction between the law governing the contract and the contract drafting and so there is no universally accepted set of rules. Determining in a particular context what approach a court will take in the event of a dispute can be a complex area open to considerable debate.[44]

It is also the case that although claims for additional costs incurred by contractors related to the extension of the completion date are commercially a related issue to the delay, as a matter of legal analysis a claim for additional money and a claim for an extension of time to complete works are different things. The rules that apply to them differ. It is common in many contracts to see a claim for an extension of time and the claim for additional money to be dealt with as separate aspects of the contract with differing rules applying to each.

[44] For a good overview and analysis of extensions of time and proof of delay in construction projects, see the Society of Construction Law Delay and Disruption Protocol (2017).

There are also circumstances where a contractor may make a claim for relief from damages or for more money outside of the contract mechanism either because the contract mechanism has broken down or because the mechanism does not provide an operative solution.

7.3. The NEC4 ECC approach to adjustment of the Completion Date

The NEC contracts have much in common with other contracts in that in addressing the question of when and how the Completion Date will be adjusted, they are seeking to provide a workable mechanism against a broader context and to achieve that the contract drafting must take into account the same considerations as other contracts.

So, we see drafting that seeks to keep claims within the four corners of the contract[45] and that covers many of the same types of event that will allow an adjustment to the Completion date, as we see in other forms of contract.

Where the NEC family differs is that the mechanisms are not simply seeking to preserve a point from which damages can be taken but to adjust the base line against which delivery will be managed. The mechanism is, therefore, one that concentrates on forecasting a future date based on the predicted effect on the work, not upon attributing liability for a known delay.

In order to predict the effect on work, the contract takes as its reference point the planned Completion date for completing the works as it is currently predicted and seeks to then adjust that date as result of the forecast of the likely effect of the event under consideration upon the current predicted date.

A forecast is only as good as the information on which it is based, and construction projects do not stand still and so the contract is seeking to achieve a mechanism where the forecast and the event are considered as soon as practicable. The outcome that the contract is aiming to achieve is an adjustment made on an agreed basis in a timely way.

The other unusual feature of the NEC family is that the time and money adjustments to the contract are made in the same mechanism and are closely related. In the NEC contract, a contractor does not prove its loss or prove the cause of the delays, but in the same process it must forecast both the time and financial implications of the event. The *Contractor* is 'compensated' through a contract mechanism that is not akin to a claim for damages (or loss and expense) and where delay is not justified by rules of causation cause and effect but by rules to establish cause and its predicted effect.

It is the Accepted Programme that gives us the base line against which to consider the effect of the event upon planned completion as predicted. It is for this reason that the Accepted Programme is fundamental to good administration of the contract.

The clause that deals with compensation events is clause 60, albeit events that are compensation events may be found in other clauses. The process relies upon a quotation being submitted by the *Contractor*, but the quotation deals with both time and money.

[45] See clause 63.6.

Clause 62.2 provides:

> Quotations for a compensation event comprise proposed changes to the Prices and any delay to the Completion Date and Key Dates assessed by the *Contractor*. The *Contractor* submits details of the assessment with each quotation. If the programme for remaining work is altered by the compensation event, the *Contractor* includes the alterations to the Accepted Programme in the quotation (emphasis added).

7.4. The alterations to the Accepted Programme

The drafting provides detail about how both the Prices are adjusted and also how delay is to be assessed. Delay is dealt with in clause 63.5, which provides:

> 63.5 A delay to the Completion Date is assessed as the length of time that, due to the compensation event, planned Completion is later than planned Completion as shown on the Accepted Programme current at the dividing date.
>
> A delay to a Key Date is assessed as the length of time that, due to the compensation event, the planned date when the Condition stated for a Key Date will be met is later than the date shown on the Accepted Programme current at the dividing date.
>
> The assessment takes into account:
>
> - any delay caused by the compensation event already in the Accepted Programme and
> - events which have happened between the date of the Accepted Programme and the dividing date.

Clause 63.1 defines the dividing date as:

> For a compensation event that arises from the *Project Manager* or the *Supervisor* giving an instruction or notification, issuing a certificate or changing an earlier decision, the dividing date is the date of that communication. For other compensation events, the dividing date is the date of the notification of the compensation event.

Other clauses give more detail about assessment:

> 63.8 The assessment of the effect of a compensation event includes risk allowances for cost and time for matters which have a significant chance of occurring and are not compensation events.
>
> 63.9 The assessment of the effect of a compensation event is based upon the assumptions that the *Contractor* reacts competently and promptly to the event and that any Defined Cost and time due to the event are reasonably incurred.

A point of general application is that like most contracts, the adjustment to the completion date is only to a later Completion Date. The Prices in the contract can be reduced for certain compensation events – see clause 60.1(1), for example – but the Completion Date can only be adjusted to a later date.[46]

From the drafting, the clause suggests several steps.

1. Identify the Accepted Programme current at the dividing date.
2. Update the programme to reflect the position at the dividing date.
3. Submit the programme submitted with the quotation.

The clause, as it is re-produced above, is the clause following a correction to earlier drafting. The correction was to modify the drafting to give additional guidance in relation to the process.

The additional drafting provides:

> The assessment takes into account
>
> - any delay caused by the compensation event already in the Accepted Programme and
> - events which have happened between the date of the Accepted Programme and the dividing date.

There are a number of ways in which extensions of time can be assessed and these are characterised in differing ways.[47] The type of analysis being undertaken here is called a 'Time Impact Analysis', the definition of which is described in the Society of Construction Law (SCL) Protocol.[48]

[46] Note acceleration under clause 33 is only by agreement. Also, there is no power to instruct the Contractor to mitigate critical delay by bringing extra resources. See the different approach in FIDIC 2017.

[47] See the following discussion of Time Impact Analysis in the Society of Construction Law Delay and Disruption Protocol (2017: p. 33): 'Criticality is determined in one of three different ways. Purely prospective critical path assessments adopt the perspective evident at the outset of the project only and take no account of progress achieved. Contemporaneous critical path assessments adopt an evolving perspective over the course of the works and take account of the effect that both historical progress and changes in the strategy for the future prosecution of the works have on predicted criticality. Retrospective critical path assessments adopt the perspective evident at the end of the project (or window of time).'

[48] See also Society of Construction Law Delay and Disruption Protocol (2017, p. 34).

Method of analysis	Analysis type	Critical path determined	Delay impact determined	Requires
Time impact analysis	Cause & effect	Contemporaneously	Prospectively	■ Logic linked baseline programme. ■ Update programmes or progress information with which to update the baseline programme. ■ Selection of delay events to be modelled.

7.5. The NEC4 ECC Practice Note 1.1 Assessing delays owing to compensation events

To supplement and support the additional drafting, a practice note was published that provides further guidance, which is re-produced in Table 3.[49]

7.6. The significance of the Practice Note

The fact that the guidance is necessary reflects the fact that there is a limit as to how far the contract drafting can go in setting out how updating the programme can be done and the differing practical considerations that will be encountered in practice.

These practice notes provide a common point of reference for the *Project Manager* and the *Contractor* in seeking to operate the process to achieve an agreed outcome.

The note reflects the fact that depending on how the process of updating the Accepted Programme has been operated, the task of updating to identify the impact of individual compensation events can be harder.

The Practice Note is seeking to provide more detail of the process to be followed to reach agreement of the alterations to the programme and the impact on planned Completion. Its relevance if agreement is not reached may be entirely hypothetical since as we will see in the absence of an agreed outcome the contract defaults to another process.

In terms of the challenges of reaching an agreed outcome, the simplest scenario is where the Accepted Programme current at the dividing date is, in fact, quite up to date.

In that case, impacting the forecast to give the alteration may be relatively straightforward.

Whether that is the case will depend upon how frequently the programme is updated and how quickly it is accepted or rejected.

The other complicating factor is whether there is a gap between the dividing date and the point at which the event began to have an effect. So, in the case where the *Contractor* notifies of an event, the dividing date could be up to 8 weeks after becoming aware of the event.

If in the intervening period the *Contractor* has updated the programme, then at the dividing date the programme should be relatively up to date.

For example, an event happens on 1 August 2022 and the *Contractor* takes 8 weeks before notifying of the event.

The dividing date could be at the end of September 2022.

If the Contract Data requires a programme to be submitted at four weekly intervals, the maximum period for which there is not a revised programme is 4 weeks.

---

[49] See: https://www.neccontract.com/getmedia/415d4555-7980-4a18-b604-eb8369f2dd64/NEC4-ECC-Practice-Note-1-1_web.pdf#:~:text=This%20practice%20note%20has%20been%20prepared

It would be expected that the programme had been updated quite close to the quotation being required.

The only reason for not having an Accepted Programme that is reasonably up to date is if the updated programme has not been submitted or the *Project Manager* has agreed to the extend time or has failed to respond.

If the *Client* sets the default interval for revising programmes to a shorter period such as bi-weekly and we have a consultative process where the *Project Manager* is involved in the process, acceptance can be very quick, and we have an Accepted Programme as our base line that will never be far from the dividing date.

A further important point is that agreement on the Completion Date is not a pre-requisite for the programme being revised.

The *Contractor* simply has to show the current Completion Date and revise its planned completion date. This means that preceding compensation events do not have to have been agreed. The programme can be constantly updated even though the Completion Date is a matter of contention. Where the planned completion date is not agreed then that will not be possible.

Again, risk whether that is ultimately included in a compensation event or is held by the *Contractor* is part of the programme. In planning its work, the *Contractor* always must allow for risk it holds. The programme is in this sense a neutral document.

If the programme is rejected, then it is because it is not a compliant programme.

In most cases there should be no difficulty on operating the process if a sensible approach to revising the programme has been adopted by the parties.[50] The Society of Construction Law Protocol describes prospective analysis as the preferred method. See Core Principal 4.

An area that is potentially a problem, albeit a problem caused by the *Project Manager* failing to follow the contract, is in the circumstance where the compensation event is one that should have been notified by the *Project Manager* but has not been.

---

[50] See Core Principal 4, and then the quote from the Society of Construction Law Delay and Disruption Protocol (2017, p. 5):
'4. Do not 'wait and see' regarding impact of delay events (contemporaneous analysis)
The parties should attempt so far as possible to deal with the time impact of Employer Risk Events as the work proceeds (both in terms of EOT and compensation).
Applications for an EOT should be made and dealt with as close in time as possible to the delay event that gives rise to the application. A 'wait and see' approach to assessing EOT is discouraged. Where the Contractor has complied with its contractual obligations regarding delay events and EOT applications, the Contractor should not be prejudiced in any dispute with the Employer as a result of the CA failing to assess EOT applications. EOT entitlement should be assessed by the CA within a reasonable time after submission of an EOT application by the Contractor. The Contractor potentially will be entitled to an EOT only for those events or causes of delay in respect of which the Employer has assumed risk and responsibility (called in the Protocol Employer Risk Events) that impact the critical path.'

Clause 61.3 provides.

61.3 The *Contractor* notifies the *Project Manager* of an event which has happened, or which is expected to happen as a compensation event if

- the *Contractor* believes that the event is a compensation event and
- the *Project Manager* has not notified the event to the *Contractor*.

If the *Contractor* does not notify a compensation event within eight weeks of becoming aware that the event has happened, the Prices, the Completion Date or a Key Date are not changed unless the event arises from the *Project Manager* or the Supervisor giving an instruction or notification, issuing a certificate or changing an earlier decision (underlining is author's own).

The underlined wording clarifies that there is no 8-week time bar where the compensation event is one which the *Project Manager* should have notified of. Instead, the position is that the *Contractor* must notify but the time is limited by clause 61.7, which provides 'A compensation event is not notified by the *Project Manager* or the *Contractor* after the issue of the Defects Certificate.'

A typical example of such a scenario is that the *Project Manager* issues information early in the contract, which is a change in Scope. The *Project Manager* fails to notify that this is a compensation event or instruct a quotation (it could be they are hoping the *Contractor* will not notice or they are unaware of the change). The *Contractor* does not pick up the change until it is reviewing as-built drawings and then puts in the notice after completion.

The date of the communication from the *Project Manager*, assuming it is an instruction in writing with a changed drawing or specification, would be the dividing date.

We now must conduct the adjustment process. This would involve adjusting the Completion Date by reference to the impact on the planned completion in the normal way by reference to the Accepted Programme current at the dividing date.

There would appear to be no difficulty with this approach, but some may consider it counter intuitive to predict the effect of something after the event. For example, see the difficulties the parties got into on *Northern Ireland Housing Executive* v. *Healthy Buildings (Ireland) Ltd* [2017] NIQB 43 a PSC contract under NEC 3 and the attitude of the court predictive assessment.

As the Completion Date only moves to a later date (except by agreement) and as each assessment is a snapshot of a prediction at a point in time, there is no conceptual or legal argument for departing from the process in this circumstance.

But the use of prospective analysis is not the general custom once disputes arise for the lawyers or the expert witnesses who are employed in the case of disputes. See, for example, the commentary in *Keating on NEC 2nd Edition*, which suggests that in principle the same rules of retrospective delay analysis apply to an NEC contract as have been applied to JCT forms of contract.[51]

[51] See *Keating on NEC Contracts 2nd Edition* (Thomas and Lee, 2022: p.443).

The argument is not a purely academic one. The question in any given case as to whether any adjustment to the completion date should be made and if so what will be answered differently depending upon whether a predictive analysis is made or a retrospective analysis.

Of course, those rules will only be tested if the parties are in dispute and as yet there have been few authorities that address the approach to delay in NEC contracts and none under the NEC4 form of contract.

The guidance note also refers to circumstances in which the parties need to either agree a position or, in default, the *Project Manager* must make their own assessment.

A first point to note is that it will not be possible to accept a quotation from the *Contractor* that does not deal with both time and money that will have the equivalent status under the contract to that which attaches to a quotation already accepted by the *Project Manager*.

A quotation dealing with both time and money is not subject to review. It is an agreed position.

But is also a complete agreement. The contract does not envisage that the financial impact is agreed but not the time impact or vice-versa.

It is not uncommon to hear of procedures where the *Project Manager* has partially accepted the assessment made by the *Contractor* in terms of money but not dealt with time.

This can be done by agreement of the Parties, but the contract does not address how that is to be treated. Further the Accepted Programme should be a compliant programme in all respects. If there are disagreements about the impact of an event and the *Project Manager's* position is that the *Contractor* has not assessed the impact correctly, then the contract defaults to using something other than the Accepted Programme.

The programme submitted for the purposes of quantifying the delay caused by a compensation event has no status, albeit the failure to submit such a programme where the event has an impact on the remaining work should disqualify the financial assessment from consideration.

We will come back to this topic in the context of the *Project Manager's* assessments in Chapter 10.

The point here is that if we are to preserve our base line of an Accepted Programme, then deals or agreements on some other or partial basis may ultimately take us away from the goal and are sub-optimal as an approach.

7.7. Examples of the adjustment process

Set out below are four examples of the scenarios envisaged by the guidance note. The first three deal with who owns the float. Example 4 considers who own the terminal float.[52]

[52] The illustrations are taken from the NEC contracts and training materials (© Copyright nec 2025).

7.7.1 Example 1 A *Client* compensation event

First, the optimum scenario is where the Accepted Programme current at the dividing date is up to date to within a few days of the dividing date and there has been no change to the actual progress that would have been expected at the dividing date. Further there have not been any other compensation events.

In the example, the compensation event is an instruction to carry out additional work.

Figure 7.1 shows the Accepted Programme current at the dividing date. The programme shows operations 1, 2 and 3 are the critical path and have a planned completion date of the end of week 7. Operations 4 and 5 are not on the critical path and would complete at the end of week 5 with 2 weeks float from the planned completion date.

Figure 7.1 The Accepted Programme current at the dividing date (Source: © Copyright nec 2025)

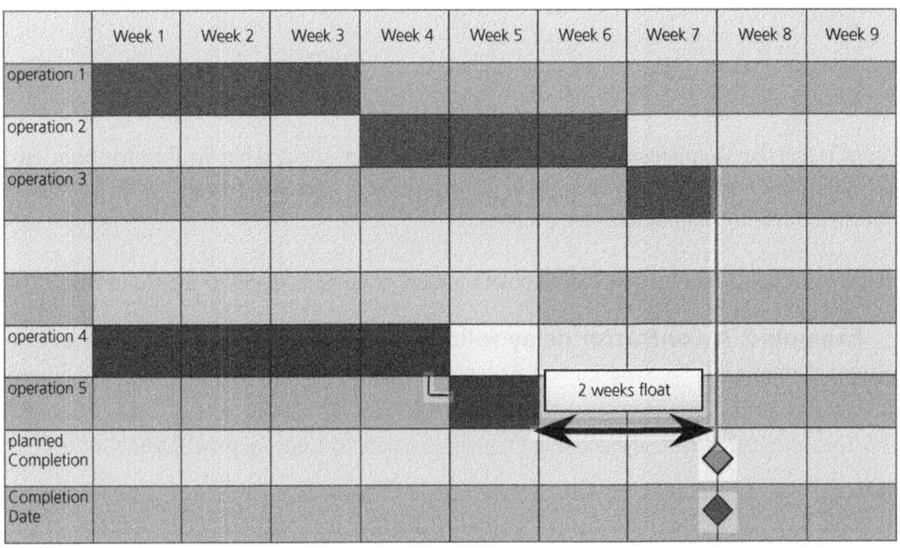

According to our facts, no other adjustments need to be made to this programme and so we can impact our compensation event.

The additional work will take 2 weeks and cannot be commenced until operation 5 has been completed. There is no impact of this work on operations 1, 2 and 3. See Figure 7.2.

Figure 7.2 Impact of compensation event (© Copyright nec 2025)

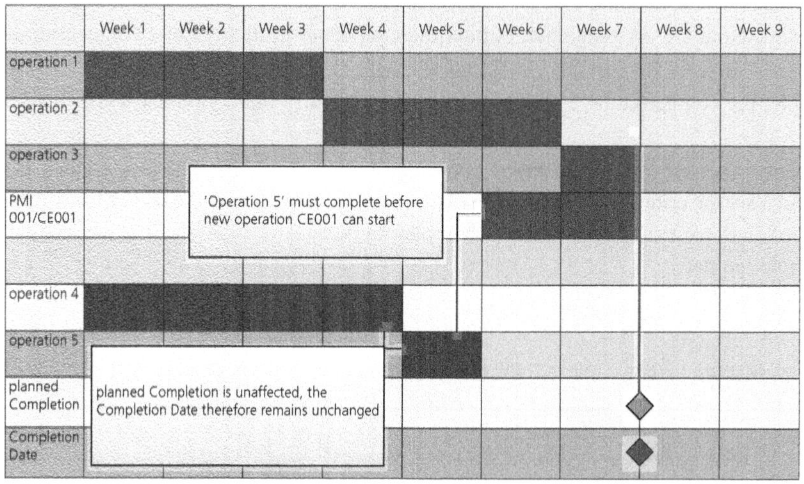

What we see is that that the planned completion date has not been affected but the float has been used up. The logic of the programme has also changed in that any delay to operations 4 and 5 or the compensation event work will now impact planned completion.

As there has been no impact on planned completion there is no adjustment to the Completion Date.

7.7.2 Example 2 A *Contractor* delay followed by a *Client* compensation event

In our second example, it is the same current Accepted Programme at the dividing date. See Figure 7.3.

Figure 7.3 The Accepted Programme current at the dividing date (© Copyright nec 2025)

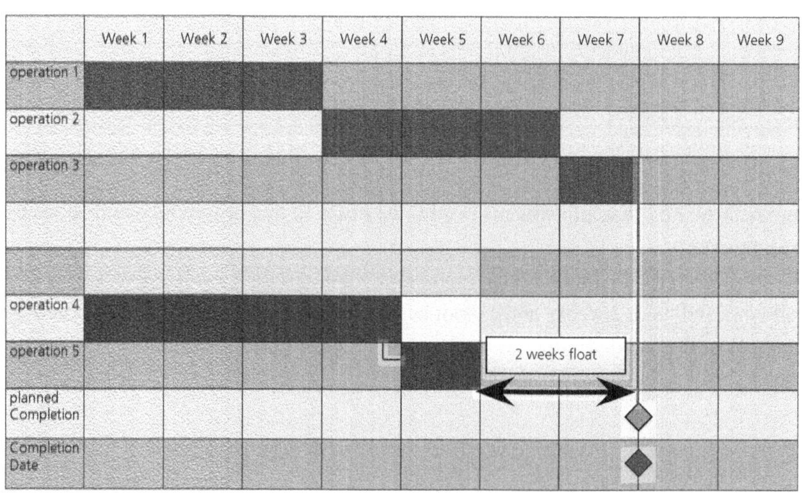

However, we know that there has been some delay to operation 4 since that programme was accepted. So, we update the programme. See Figure 7.4.

Figure 7.4 Taking into account progress against activities (© Copyright nec 2025)

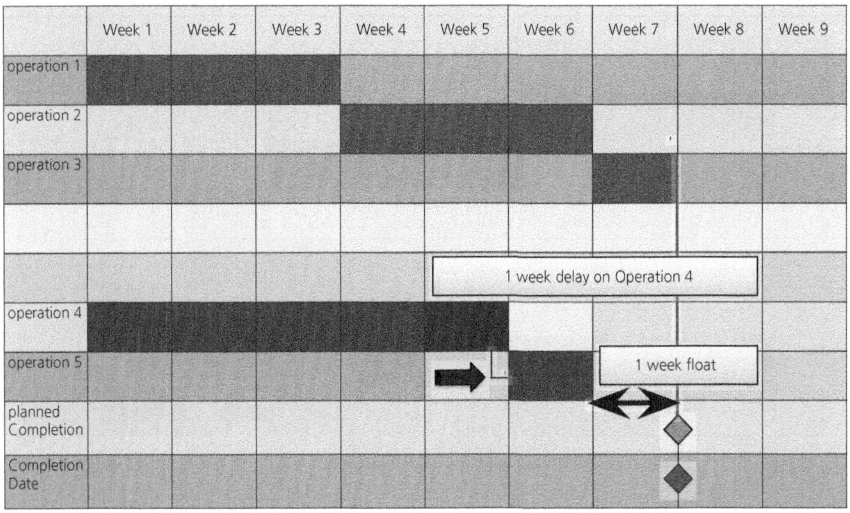

The float has been reduced from 2 weeks to 1 week. We now impact the compensation event. See Figure 7.5.

Figure 7.5 Adding the compensation event (© Copyright nec 2025)

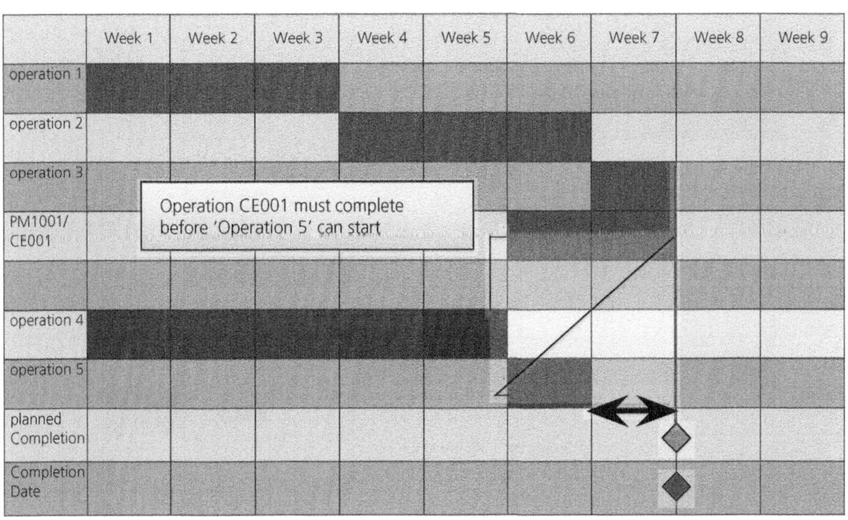

The effect of the compensation event is to add 2 weeks to completion of work for operations 4 and 5 which and operation 5 cannot start until this work is complete and will now be delayed by 2 weeks. See Figure 7.6.

Figure 7.6 Impact of the compensation event (© Copyright nec 2025)

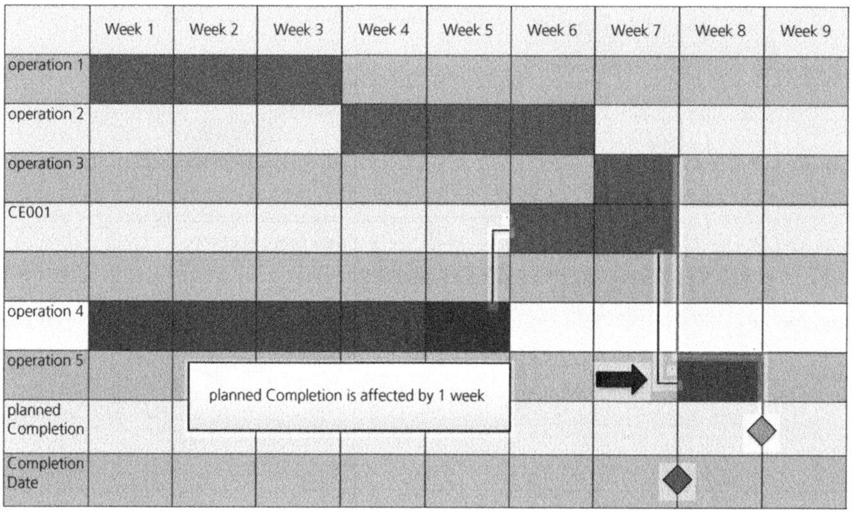

Planned completion has been moved by 1 week from end of week 7 to the end of week 8 and so the Completion Date will also have to move by 1 week. See Figure 7.7 below.

Figure 7.7 Change to the Completion Date (© Copyright nec 2025)

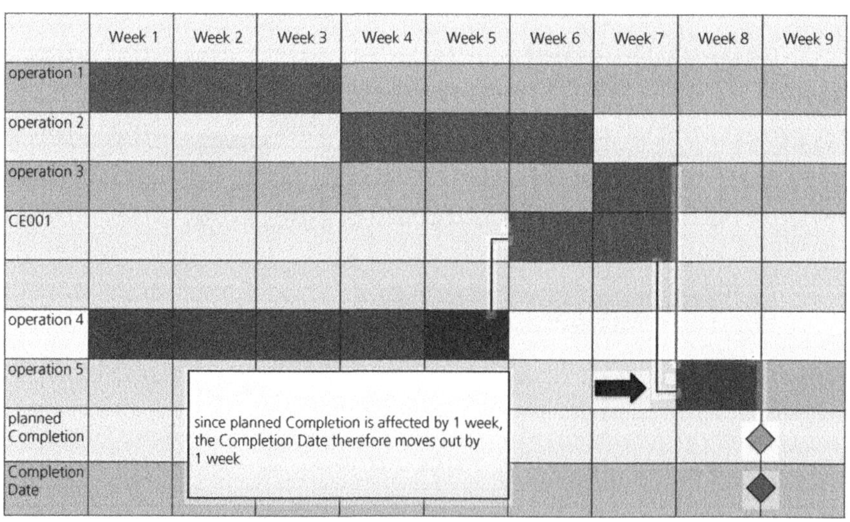

7.7.3 Example 3 *Contractor* improved progress followed by *Client* compensation event

In our next example, we have the same programme as the Accepted Programme current at the dividing date. See Figure 7.8.

Figure 7.8 The Accepted Programme current at the dividing date

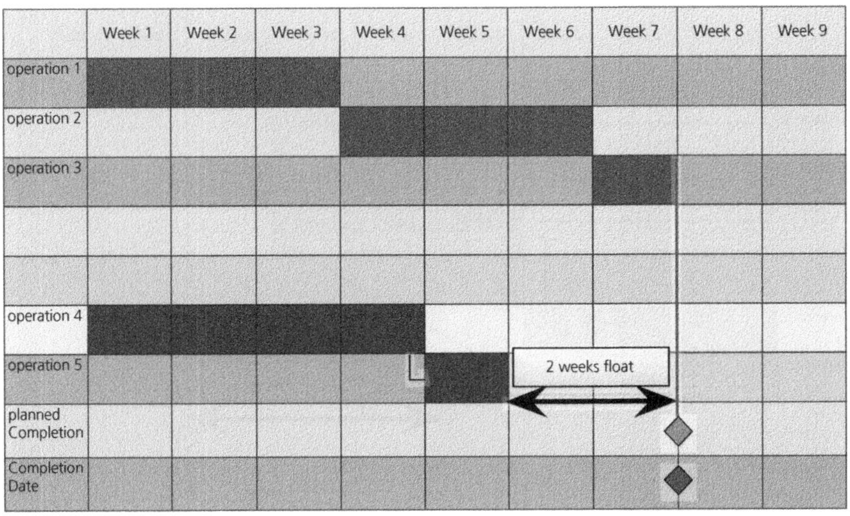

Since that programme was accepted, the *Contractor* has made better progress on operation 4 and is 1 week ahead of programme. The programme is updated. See Figure 7.9.

Figure 7.9 Taking into account progress against activities (© Copyright nec 2025)

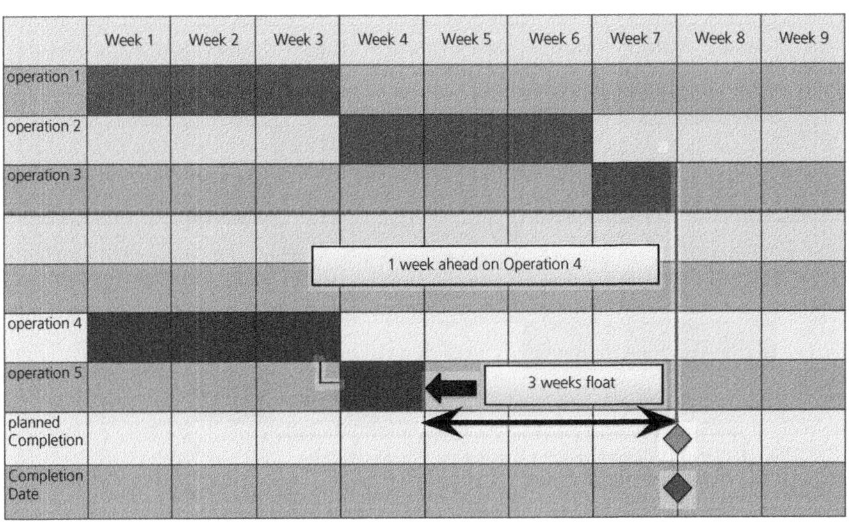

The compensation event is additional work that will take 3 weeks. It cannot start until operation 4 is complete and operation 5 cannot start until the additional work is complete. Introduce the CE 001. See Figure 7.10.

Fig 7.10 Adding the compensation event (© Copyright nec 2025)

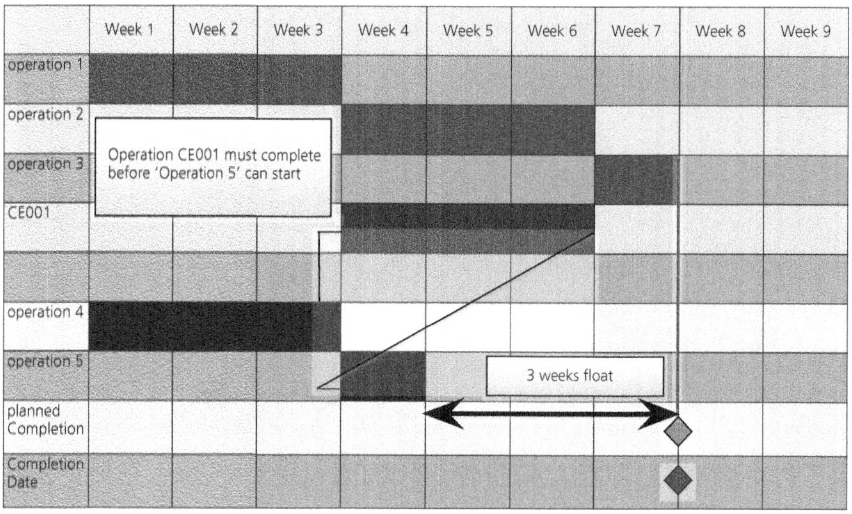

The effect of the compensation event is to use all the existing float, but it has no effect on planned completion. There is no adjustment to the Completion Date. See Figure 7.11.

Figure 7.11 Impact of the compensation event (© Copyright nec 2025)

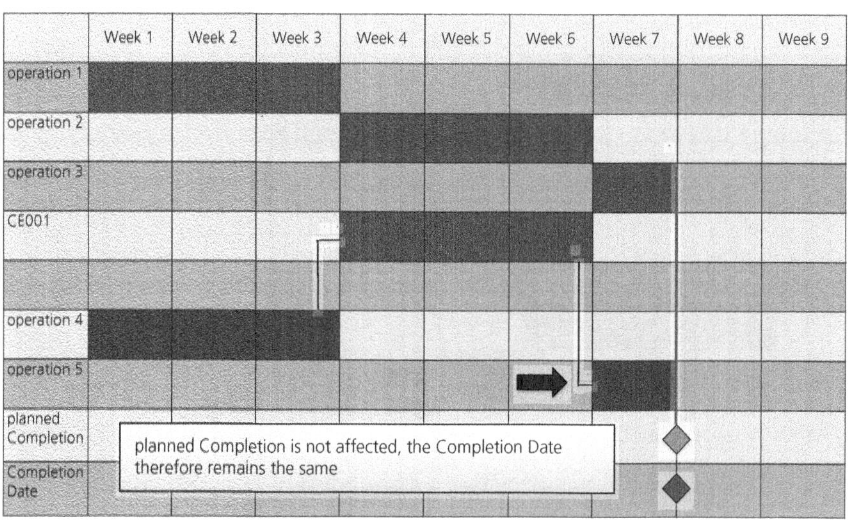

7.7.4 Comments on Examples 1–3

These examples are all based on a compensation event or delay that did not directly effect the critical path. The examples concern who owns the float in the programme and as demonstrated it is owned by the person who uses it first. That highlights the importance of the programme accurately representing the true position at the point the compensation event is impacted.

The next example will look at how the contract deals with a different scenario which is where the impact of the compensation event is on the critical path and the question of who owns the terminal float.

7.8. Terminal float

7.8.1 Terminal float – Example 4

The programme Figure 7.12 shows that the *Contractor's* planned completion date is the end of week 5. The Completion Date is the end of week 8. The period between planned completion and the Completion Date is called the 'terminal float'. This belongs to the *Contractor*. The reason being is that the reference point for the adjustment completion date is:

'63.5 A delay to the Completion Date is assessed as the length of time that, due to the compensation event, planned Completion is later than planned Completion as shown on the Accepted Programme current at the dividing date (as adjusted).'

Figure 7.12 The 'terminal float' (© Copyright nec 2025)

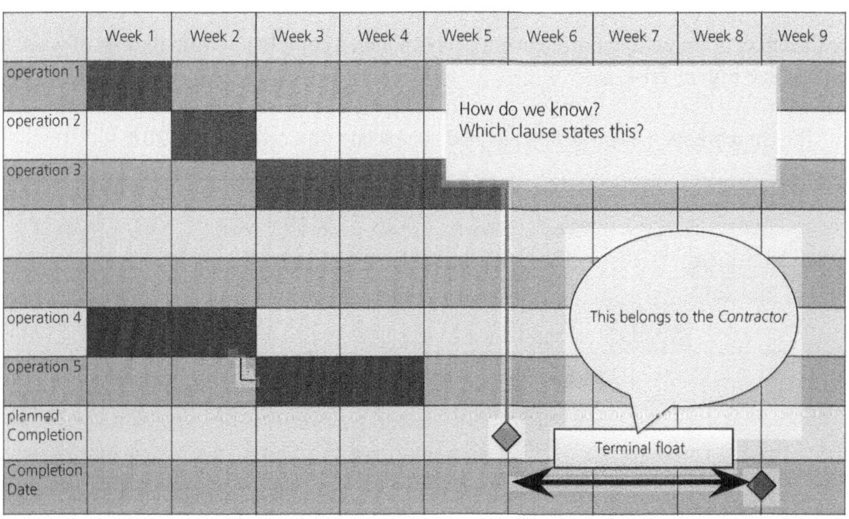

We now introduce a compensation event that delays or increases the work on the critical path that will feed through to a later planned completion date. See Figure 7.13. The event is additional work that takes 1 week and operation 3 cannot start until it is completed, so it impacts operation 3 on the critical path.

Figure 7.13 A compensation event delays or increases the work on the critical path (© Copyright nec 2025)

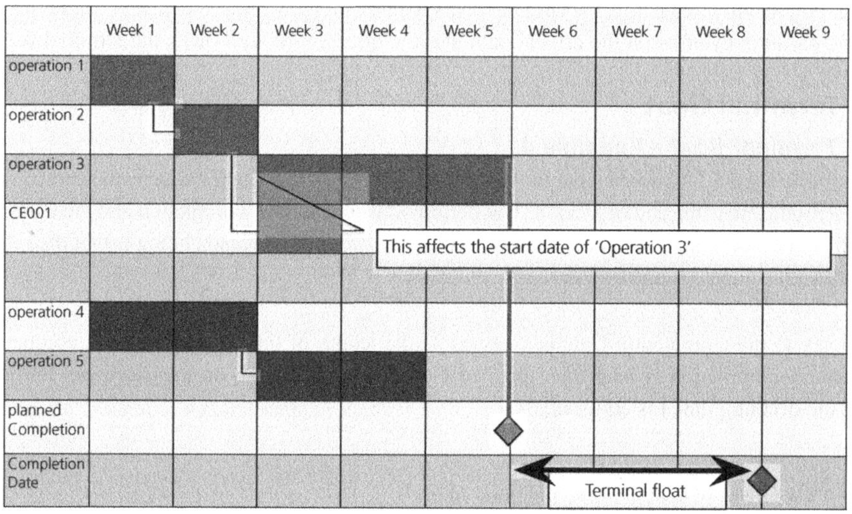

We can see the effect of the compensation event delays planned completion from the end of week 5 to the end of week 6. See Figure 7.14.

Figure 7.14 The compensation event delays planned completion (© Copyright nec 2025)

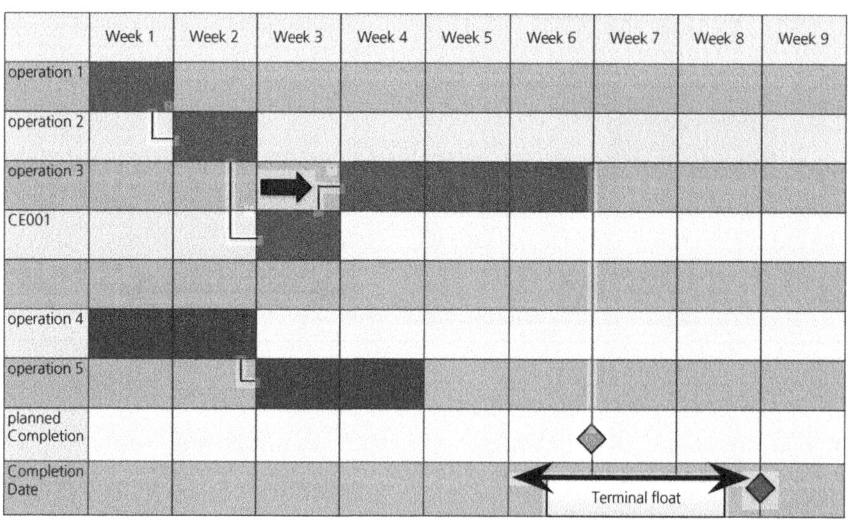

Because we have moved planned completion by 1 week, we adjust the completion date by 1 week. So, we move the completion date from the end of week 8 to the end of week 9 and the terminal float is preserved. See Figure 7.15.

Figure 7.15 The completion date is adjusted (© Copyright nec 2025)

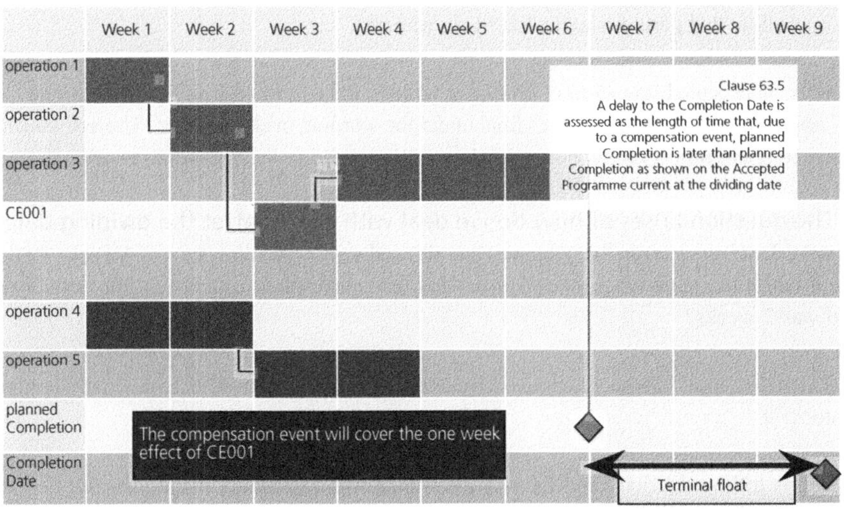

7.8.2 The implications of the drafting

From the preceding examples, it can be seen that the wording that governs the basis for an extension has some fundamental implications.

First, the Completion Date is adjusted if, according to a specific analysis at a specific point in time, the forecast date for planned completion is affected by the compensation event.

The assessment is made by reference to the dividing date and we have actual information on resources and the occurrence of events up to the dividing date and a forecast after the dividing date.

The dividing date is itself governed by the point at which the compensation event is either notified by the *Contractor* or in the case of a compensation event that arises from the *Project Manager* or the *Supervisor* giving an instruction or notification, issuing a certificate or changing and earlier decision, the dividing date is the date of that communication.

This will dictate the Accepted Programme that we use, which is the one current at the dividing date, but[53] there are additional requirements to consider in making the assessment:

- *any delay caused by the compensation event already in the Accepted Programme*
- *events which have happened between the date of the Accepted Programme and the dividing date.*

[53] This programme is not necessarily the Accepted Programme as there is only one Accepted Programme at any time, which supersedes the previous Accepted Programmes. So, it applies if the Accepted Programme was not the Accepted Programme at the dividing date.

The drafting does not invite us to consider what has happened between the dividing date and the point in time at which the quotation is accepted or rejected.

In a typical case where the *Project Manager* has issued an instruction, the dividing date will be the date of the instruction, at the same time the *Contractor* will be instructed to provide the quotation, there will be 3 weeks to provide the quotation and the *Project Manager* will have 2 weeks to accept or not accept the quotation (unless that period is extended by agreement).

If the quotation is accepted, the *Project Manager* notifies the *Contractor* and that constitutes 'implementation' under clause 66.1 and becomes final under the wording in clause 66.3: 'The assessment of an implemented compensation event is not revised except as stated in these conditions of contract.'[54]

7.8.2.1 The question arises of how do we deal with events after the dividing date?

If, for example, the *Contractor* in the quotation has allowed a time risk allowance of 4 weeks for delivery of a material but at the point where the *Project Manager* makes their assessment the delivery date is known and was 2 weeks?

The shorter delivery time has been achieved by some good fortune in finding a small supplier with available stock.

If the drafting is followed strictly and the risk allowance was reasonably made then the *Contractor's* quotation should be accepted. If the same issue is considered from the other perspective and the allowance was for 4 weeks but having submitted it, and before the quotation was accepted, the *Contractor* realises it made a mistake and the material will take 6 weeks to be delivered, there is no drafting that allows the *Contractor* to revise its quotation.

It suggested that some *Project Managers,* in circumstance where the *Contractor* benefits, might simply not accept the *Contractor's* quotation but that would not be justified by the drafting which limits the *Project Manager* to not accepting the quotation and making their own assessment 'if the *Project Manager* decides that the *Contractor* has not assessed the compensation event correctly'.

In our example, so long as the original forecast was reasonable there would not have been anything 'incorrect' in the quotation.

Another suggestion is the *Project Manager* instructs the *Contractor* to submit a revised quotation. But again, the reasons have to be stated and could be subject to review.

Further, if that course were adopted it would be contrary to the spirit of mutual trust and co-operation if that was done in a partisan way such that it was only where the *Client* benefited that the quotation was 'revised'.

Arguably, any agreement to take into account events after the dividing date would also be a change to the contract and would need to be confirmed by the Parties under clause 12.3.

[54] Potentially, it might be disputed in Adjudication; see Chapter 8.

7.8.2.2 Other implications that follow from the drafting

If there is float in the programme, as we have seen, it gets used by whoever gets there first; in the sense that once the float is gone it cannot be used by the other party.

If the *Contractor* has not updated a programme to show actual progress the programme is non-compliant. The *Project Manager* is perfectly entitled to not accept it.

Terminal float is also dictated by actual progress and planned completion, so the *Contractor* cannot retain terminal float if the reality is that planned completion is delayed. The only way to retain the float is to get an extension to the Completion Date.

Planned Completion is a point in time that should be dictated by the logic of the programme and the progress of work. It does not matter what drives planned Completion whether that is a compensation event or contractor delay, planned Completion should be an objectively discernible point in time which the *Project Manager* has either to accept the programme, assuming the programme is correctly produced and updated or to not accept the programme, if it is not.

7.8.2.3 Identifying the cause of delay

The NEC 4 Practice Note states:

> 'If multiple compensation events arise the assessments should be carried out sequentially, separately and prospectively. However, when a number of events arise over a short period of time, it may be sensible, by agreement, to consider them together. The impact of other compensation events existing prior to the dividing date should be recognised before making the new compensation event assessment.'

This is consistent with the approach taken in the SCL Delay Protocol but, as we shall see below, the assessment of multiple compensation events can of itself raise several questions.

The starting point is first to consider that the programme is a dynamic model that changes as the project proceeds to reflect how the project is progressing. When we assess a compensation event, we are taking a snapshot of the project and using the software to model to our predicted planned completion date.

If the progress reporting on the contract is good such that the snapshot at the 'dividing date' is accurate, then the major challenge is to distinguish which change is owing to the compensation event and which change is just change that is the *Contractor's* risk.

The burden of proof is placed on the *Contractor* to identify what delay is owing to the compensation event. Again, this takes us back to the progress reporting. It will not be sufficient to show that operations completed later than anticipated. The resources that have been impacted either in terms of changes to resources or the under-utilisation of resources must be shown. Also, the prediction of the delay must also assume how future operations will be impacted. This again takes us back to resources and productivity or programme logic.

The supporting information for the time adjustment will be closely linked to the justification for any financial compensation that is to be claimed. The basis for the claim is Defined Cost, which is a resource-based valuation.[55]

---

[55] NEC4 ECC Defined Cost is set out in the relevant Schedule of Cost Components.

In the *CIOB Guide to Good Practice in the Management of Time in Major Projects* (paragraph 4.1.4) the point is made that programmes are produced at differing levels of detail at different stages. The CIOB planning protocol and other guidance on the production of programmes suggest that for the construction phase of the project the programming must be at a significant level of detail. For actual progress reporting the resources used should be known. The programme should be in sufficient detail to understand the resources that were assumed and the productivity that was assumed. There is no standard convention for this.

Clause 32.1 of the NEC4 ECC sets out requirements that all operations must be programmed and further that the *Contractor* must provide a statement of how the *Contractor* plans to do the work and to identify the principal Equipment and resources that will be used.

It is also important to have some standardisation of reporting and identification of resources that will allow the resources to be related to the relevant operations. *The CIOB Guide to Good Practice in the Management of Time in Major Projects* provides guidance on structuring of the programme in work breakdown structures and activity identification at differing levels of detail depending on the purpose for which the programme is to be used. These codes can be used also to match resource records.[56]

There is anecdotal evidence that *Contractor's* resist programming at this level of detail.[57] Further, that detailed resourcing information may not be available for the whole of a project in its early stages and so the *Contractor* may struggle to provide this information in a first programme.

The CIOB guide envisages that the programme is a rolling programme that updates the resourcing content in a wave such that the high-density content covers the next three months. However, the Code does not assume that any prediction of delay taken from the programme will be fixed and so if assumptions or resourcing later in the programme change when updated so as to reduce or increase the impact of an event the delay can be re-calculated.

The NEC contracts only permits this in very limited circumstances.[58]

The rolling wave suggestion may not be the answer but if that is made a requirement in the Scope for updating programmes, generally there may also need to be further drafting to make this a requirement when submitting a quotation.

Under the standard drafting, the *Project Manager* should be confident that even though the *Contractor* may suggest it will provide more detail in resource terms for later parts of the programme at a later stage, there is sufficient clarity in any Accepted Programme to identify the resource assumptions and methods that drive the logic and durations, time risk allowances in the programme. Failure to do so could make identification of delay and impacts upon planned completion difficult leading to loss of the Accepted Programme.

---

[56] See Part 4 of the *Guide to Good Practice in the Management of Time in Major Projects 2nd Edition* (CIOB, 2018).

[57] See Hughes (2022: p. 202) in *NEC4 A Guide for Better Delivery of Projects in the Building Sector.*

[58] See clause 61.6 and clause 60.1(17).

7.9. Multiple compensation events

The challenge for multiple compensation events is that each event would normally be assessed individually and in the order in which they impact the works. Contemporaneous critical path assessments adopt an evolving perspective over the course of the works and take account of the effect that both historical progress and changes in the strategy for the future prosecution of the works have on predicted criticality.[59]

But when the assessment is to be made and which programme will be the basis for the assessment is driven by the contract. See clause 65.

> 63.5 A delay to the Completion Date is assessed as the length of time that, due to the compensation event, planned Completion is later than planned Completion as shown on the Accepted Programme current at the dividing date (underlining is author's own).

The dividing date is dictated by who notified of the event see clause 63.1:

> For a compensation event that arises from the *Project Manager* or the *Supervisor* giving an instruction or notification, issuing a certificate or changing an earlier decision, the dividing date is the date of that communication. For other compensation events, the dividing date is the date of the notification of the compensation event.

It is quite possible that the dividing date for an event notified by the *Contractor* is some time after the event impacted the works and that in the intervening period a later event notified by the *Project Manager* has impacted the works but has an earlier dividing date.

Let us look at an example.

7.9.1 Example 5 Part 1: Identifying the first event to assess the importance of Notices and the dividing date

Let us assume there are two compensation events.

The first event to impact the works is 'unforeseen physical conditions'; the second event is an instruction by the *Project Manager* to stop work.

The unforeseen conditions are first encountered at the start of week 3. The *Contractor* takes some time to investigate them and then notifies of a compensation event.

The notification of unforeseen physical conditions is given at the end of week 6.

The dividing date for this event is the date of the notification, which is the end of week 6.

---

59 See the Society of Construction Law Delay and Disruption Protocol (2017: paras 4.7–4.10).

The *Contractor* must submit its quotation by the end of week 10.[60]

Let us now assume that at the start of week 4 the *Project Manager* instructs the *Contractor* to stop work owing to problems with the re-location of the materials stored by the *Client* on an area of the site.

The *Project Manager's* instruction ('PMI') is a compensation event under clause 60.1(4).

The PMI is the first compensation event notified and is CE1 and the dividing date for this compensation event is the start of week 4.

The quotation should be provided by the start of week 7.

The quotation for CE1 must be provided 4 weeks before the quotation for CE2.

But the compensation events are to be assessed in the reverse order to that which the events impacted the works.

Also, for each event we must base the assessment on the Accepted Programme current at the dividing date.

Let us make some more assumptions in our example and see how the assessment process works.

First, we need a programme.

The works are construction of a new warehouse on the site of a factory. It is a steel frame building on concrete slab with some plant to maintain temperature/humidity levels required for storage.

The *Contractor's* programme was contained in Contract Data part two at the tender stage. See Figure 7.16 Tender Programme.

The contract requires the programme to be revised at an interval of 4 weeks.

Access dates were given in the tender.

An area of the site contains some materials for production that the *Client* is re-locating. The *Client* has identified Area A as not impacted by this and Area B that covers the effected land.

The *Client* has given an *access date* of the start of week 1 for Area A.

The *Client* has given an *access date* of the start of week 3 for Area B.

---

[60] The *Project Manager* has 1 week to reply to the notification and either accept that the notified event is a compensation event or not. If it is a compensation event, the *Project Manager* instructs the *contractor* to provide a quotation. (This period can be extended by agreement).

The *Contractor's* programme showed that access for Area A for site clearance would be taken at the start of week 1 and the operation continues until the start of week 6.

Access to Area B would be taken at the start of week 3 and the operation continues until the start of week 6.

At the start of week six, excavation would commence in both Areas A and B. The excavation is programmed to continue in both areas the until end of week 7. The excavation is to the same depth across Areas A and B, and it is not practical to excavate just to the boundary of Area B.

The excavation cannot start until both Areas A and B have been cleared.

The programme shows the operations following excavation. The *Client* has stated in the Scope commissioning of the plant must be complete for Completion and the last activity is the commissioning.

The planned Completion date is shown as the end of week 20, Friday 19 May 2023.

The Completion Date is the end of week 21, Friday 26 May 2023.

Operations outlined in red (digital version) or dark grey (print version) are on the critical path. The site clearance of both Areas A and B are on the critical path as is the following excavation.

Let us introduce the two compensation events.

The first event to impact the works is 'unforeseen physical conditions'. The second event is an instruction by the *Project Manager* to stop work.

The unforeseen conditions are first encountered in the start of week 3 in Area A. The *Contractor* takes some time to investigate them and then notifies of a compensation event under clause 60.1(12).

The notification of unforeseen physical conditions is given at the end of week 6.

The dividing date for this event is the date of the notification, which is the end of week 6

The *Contractor* must submit its quotation by the end of week 10.[61]

[61] The *Project Manager* has 1 week to reply to the notification and either accept that the notified event is a compensation event or not. If it is a compensation event, the *Project Manager* instructs the contractor to provide a quotation. (This period can be extended by agreement).
The Contractor would start their quotation at the end of week 7 and has 3 weeks to provide it which would take us to end of week 10.
The *Project Manager* would have to accept the quotation within 2 weeks or make their own assessment. That takes us to the end of week 12.

Let us now assume that at the start of week 4 the *Project Manager* instructs the *Contractor* to stop work in Area B owing to problems with the re-location of the materials. The instruction is to stop work for 2 weeks in Area B.

The *Project Manager's* instruction ('PMI') is a compensation event under clause 60.1(4).

The PMI is the first compensation event notified and is CE1 and the dividing date for this compensation event is the start of week 4.

The quotation should be provided by the start of week 7.

A revised programme was submitted at the end of week four. The *Project Manager* reviewed this programme, and it was accepted at the start of week 5 and is up to date to the end of week 4. This programme is not an assessment of the compensation event.

In our example, there is some tension between the assessment timetable and the order in which the events occurred. Does this make any difference? Could we assess CE1 first or does that give a wrong answer? Set out below is the working of the assessment.

7.9.2 Example 5 Part 2: The Assessment of the CE1 instruction to stop work

For this compensation event, the tender programme (Figure 7. 16) is the current Accepted Programme at the dividing date the start of week 4.

Figure 7.16 Tender Programme (John Marshall of Fermata Ltd.)

Line	Name	Duration	Start	Finish	Percent complete
1	Access to Area A		02/01/2023	02/01/2023	
2	Site clearance to Area A	5w	02/01/2023	03/02/2023	
3	Access to Area B		16/01/2023	16/01/2023	
4	Site clearance to Area B	3w	16/01/2023	03/02/2023	
5	Excavation for base slab	2w	06/02/2023	17/02/2023	
6	Construct base slab	3w	20/02/2023	10/03/2023	
7	Erect steel frame	3w	13/03/2023	31/03/2023	
8	Install roof sheeting, cladding and insulation	3w	03/04/2023	21/04/2023	
9	Install mechanical and electrical equipment	2w	24/04/2023	05/05/2023	
10	Commission	2w	08/05/2023	19/05/2023	
11	Planned completion		19/05/2023	19/05/2023	
12	Completion		26/05/2023	26/05/2023	

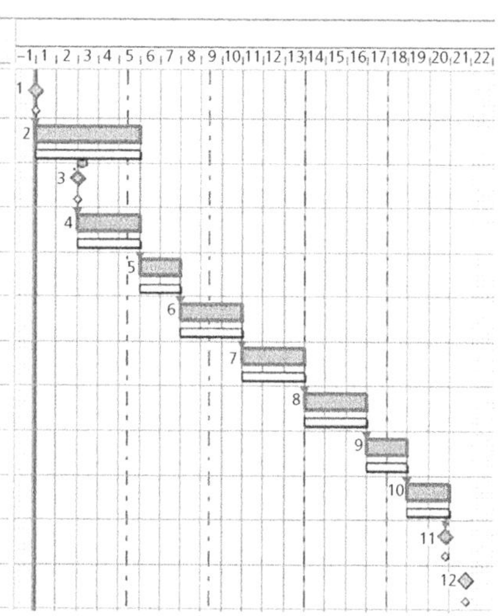

To make our assessment, the current Accepted Programme at the dividing date needs to be updated to the dividing date start of week 4. This update is not necessarily a fully revised programme, it is not a revision to the Accepted Programme under clause 32.1 but is simply the assessment under clauses 62.2 and 63.5 and so it is sometimes referred to as a 'fragnet' or subnet and can be an extract of the Accepted Programme.

The assessment must take into account

- any delay caused by the compensation event already in the Accepted Programme and

It must also show

- events which have happened between the date of the Accepted Programme and the dividing date.

For the example here the whole programme has been re-produced to show the impact analysis. See Figure 7.17.

Figure 7.17 Tender programme updated to start of week 4 (John Marshall of Fermata Ltd.)

Line	Name	Duration	Start	Finish	Percent complete
1	Access to Area A		02/01/2023	02/01/2023	100.00
2	Site clearance to Area A	8w	02/01/2023	03/03/2023	40.00
3	Access to Area B		16/01/2023	16/01/2023	100.00
4	Site clearance to Area B	3w	16/01/2023	03/02/2023	34.00
5	Excavation for base slab	2w	06/03/2023	17/03/2023	
6	Construct base slab	3w	20/03/2023	07/04/2023	
7	Erect steel frame	3w	10/04/2023	28/04/2023	
8	Install roof sheeting, cladding and insulation	3w	01/05/2023	19/05/2023	
9	Install mechanical and electrical equipment	2w	22/05/2023	02/06/2023	
10	Commission	2w	05/06/2023	16/06/2023	
11	Planned completion		19/06/2023	19/06/2023	

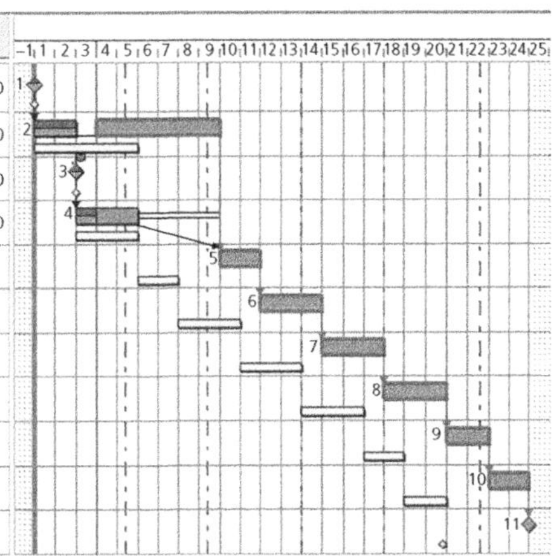

Red (digital version) and grey (print version) = actual progress

Red outline (digital version) and grey (print version) = operations on the critical path

At the dividing date there is 1 week delay up to the start of week 4 caused by CE2, the physical conditions to the site clearance in Area A. Site clearance in Area A should have been 60 % complete but is only 40 % complete. At this time the *Contractor* is predicting further delay owing to CE2, the physical conditions during the excavation that impacts excavation of Area A, delaying its completion by 3 more weeks owing to lower productivity in the excavation than anticipated.

The programme now shows planned completion as 4 weeks later than the tender programme and is now the end of week 24, Friday 16 June 2023.

At the dividing date week start of week 4, CE1 has been issued. The impact of CE1 is 2 weeks predicted delay.

The effect of the CE1 would be separated from the delay being caused by CE2 the physical conditions by the drafting that states a delay to the Completion Date *is assessed as the length of time that, due to the compensation event, planned Completion is later than planned Completion as shown on the Accepted Programme current at the dividing date (as modified by the fragnet).*

This can be achieved by the coding of resources or changes to logic or sequence by reference to the event which had caused the change. CE1 is colour coded in lighter green. See Figure 7.18, digital version.

Figure 7.18 CE1 identified (John Marshall of Fermata Ltd.)

Line	Name	Duration	Start	Finish	Percent complete
1	Access to Area A		02/01/2023	02/01/2023	100.00
2	Site clearance to Area A	8w	02/01/2023	03/03/2023	40.00
3	Access to Area B		16/01/2023	16/01/2023	100.00
4	CE1 - stop work for two weeks	2w	23/01/2023	03/02/2023	
5	Site clearance to Area B	3w	16/01/2023	17/02/2023	34.00
6	Excavation for base slab	2w	06/03/2023	17/03/2023	
7	Construct base slab	3w	20/03/2023	07/04/2023	
8	Erect steel frame	3w	10/04/2023	28/04/2023	
9	Install roof sheeting, cladding and insulation	3w	01/05/2023	19/05/2023	
10	Install mechanical and electrical equipment	2w	22/05/2023	02/06/2023	
11	Commission	2w	05/06/2023	16/06/2023	
12	Planned completion		19/06/2023	19/06/2023	

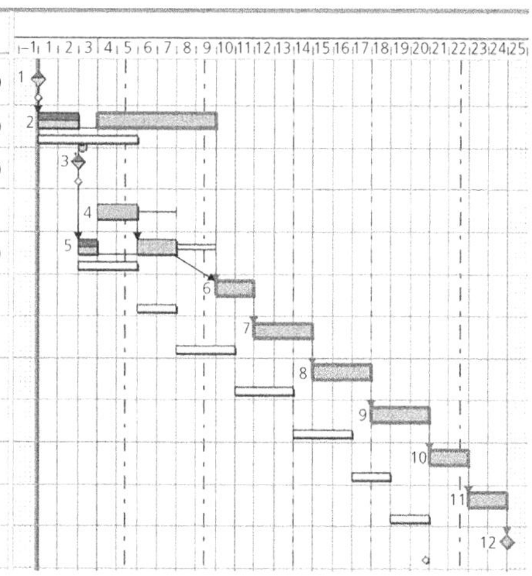

The PMI has no effect on planned Completion. The critical path has changed, and Area B has 2 weeks float because 3 weeks anticipated delay in Area A has been included in the assessment. As the PMI has not had any effect on planned Completion there would be no adjustment to the Completion Date.

7.9.3 Example 5 Part 3: Assessment of the CE2 unforeseen physical conditions

CE2 has a dividing date of the end of week 6 (the date of the notification).

A revised programme was submitted at the end of week 4. The *Project Manager* reviewed this programme, and it was accepted at the start of week 5 and is up to date to the end of week 4.

The current Accepted Programme at the dividing date is the revised programme. See Figure 7.19.

Figure 7.19 Tender programme updated to start of week 4 (John Marshall of Fermata Ltd.)

Line	Name	Duration	Start	Finish	Percent complete
1	Access to Area A		02/01/2023	02/01/2023	100.00
2	Site clearance to Area A	8w	02/01/2023	03/03/2023	50.00
3	Access to Area B		16/01/2023	16/01/2023	100.00
4	CE1 - stop work for two weeks	2w	23/01/2023	03/03/2023	50.00
5	Site clearance to Area B	3w	16/01/2023	17/02/2023	34.00
6	Excavation for base slab	2w	06/03/2023	17/03/2023	
7	Construct base slab	3w	20/03/2023	07/04/2023	
8	Erect steel frame	3w	10/04/2023	28/04/2023	
9	Install roof sheeting, cladding and insulation	3w	01/05/2023	19/05/2023	
10	Install mechanical and electrical equipment	2w	22/05/2023	02/06/2023	
11	Commission	2w	05/06/2023	16/06/2023	
12	Planned completion		19/06/2023	19/06/2023	
13	Completion		26/05/2023	26/05/2023	

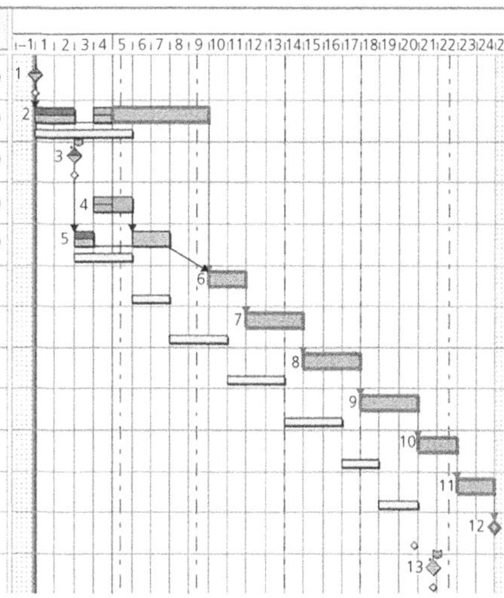

This is the update to the tender programme. It would show any actual or predicted delay in week 3 to the end of end week 4 owing to CE2 the physical conditions or other events. See clause 32.1.[62]

[62] Clause 32.1 The Contractor shows on each revised programme

- the actual progress achieved on each operation and its effect upon the timing of the remaining work,
- how the Contractor plans to deal with any delays and to correct notified Defects and
- any other changes which the Contractor proposes to make to the Accepted Programme.

The current Accepted Programme at the dividing date 7.19 then needs updating as the programme submitted with the quotation. See Figure 7.20.

This is the assessment, and it is not just a revision of the current Accepted Programme; it is sometimes referred to as a 'fragnet' or an extract of the current Accepted Programme. Clause 65.3 further identifies what must be taken into account.

Clause 63.5

The assessment takes into account.

- *any delay caused by the compensation event already in the Accepted Programme and*

And must also show.

- *events which have happened between the date of the Accepted Programme and the dividing date.*

Taking the initial bullet point first.

Figure 7.19 already shows all the delays in the Accepted Programme for CE2 and any other effects up to the end of week four. The update must now show any events between the end of week 4 and the end of week 6.

The extract of the programme will capture the actual effect of the events that have happened up to the end of week 6 of the dividing date.

The PMI CE1 was issued at the start of week 4 and so if that had any actual impact on the work at the dividing date at the end of week 6, that would be shown. For the reasons explained, that would also have to factor in the forecast effects of CE1 on planned completion.

Also captured in the revised extract will be the actual delay caused by CE2 in terms of the delay to the activities and any actual effect on the progress of work before the dividing date (week 6) and in addition there would be a forecast of its further effect.

The effect of the CE1 would be separated from the delay being caused by CE2 by the drafting that refers us to assessing. '*A delay to the Completion Date is assessed as the length of time that, due to the compensation event, planned Completion is later than planned Completion as shown on the Accepted Programme current at the dividing date.*'

Unless there was some further change to progress or the predicted effect of any event between the start of week 4 and the end of week 6 (the dividing date) then the extract would be a repeat of the position shown on the revised programme in Figure 7.19. See Figure 7.20.

Figure 7.20 Programme updated to end of week 6 (John Marshall of Fermata Ltd.)

Line	Name	Duration	Start	Finish	Percent complete
1	Access to Area A		02/01/2023	02/01/2023	100.00
2	Site clearance to Area A	8w	02/01/2023	03/03/2023	70.00
3	Access to Area B		16/01/2023	16/01/2023	100.00
4	CE1 - stop work for two weeks	2w	23/01/2023	03/02/2023	100.00
5	Site clearance to Area B	3w	16/01/2023	17/02/2023	67.00
6	Excavation for base slab	2w	06/03/2023	17/03/2023	
7	Construct base slab	3w	20/03/2023	07/04/2023	
8	Erect steel frame	3w	10/04/2023	28/04/2023	
9	Install roof sheeting, cladding and insulation	3w	01/05/2023	19/05/2023	
10	Install mechanical and electrical equipment	2w	22/05/2023	02/06/2023	
11	Commission	2w	05/06/2023	16/06/2023	
12	Planned completion		19/06/2023	19/06/2023	

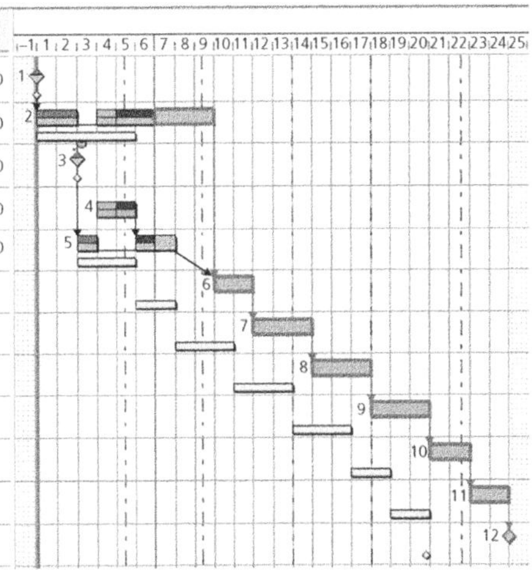

To identify the delay caused by the past and future effect of CE2 that is now introduced, and colour coded in light green (see digital version, Figure 7.21).

Figure 7.21 CE2 identified (John Marshall of Fermata Ltd.)

Line	Name	Duration	Start	Finish	Percent complete
1	Access to Area A		02/01/2023	02/01/2023	100.00
2	CE2 - unforeseen ground conditions	4w	06/02/2023	03/03/2023	
3	Site clearance to Area A	8w	02/01/2023	03/03/2023	70.00
4	Access to Area B		16/01/2023	16/01/2023	100.00
5	CE1 - stop work for two weeks	2w	23/01/2023	03/02/2023	100.00
6	Site clearance to Area B	3w	16/01/2023	17/02/2023	67.00
7	Excavation for base slab	2w	06/03/2023	17/03/2023	
8	Construct base slab	3w	20/03/2023	07/04/2023	
9	Erect steel frame	3w	10/04/2023	28/04/2023	
10	Install roof sheeting, cladding and insulation	3w	01/05/2023	19/05/2023	
11	Install mechanical and electrical equipment	2w	22/05/2023	02/06/2023	
12	Commission	2w	05/06/2023	16/06/2023	
13	Planned completion		19/06/2023	19/06/2023	

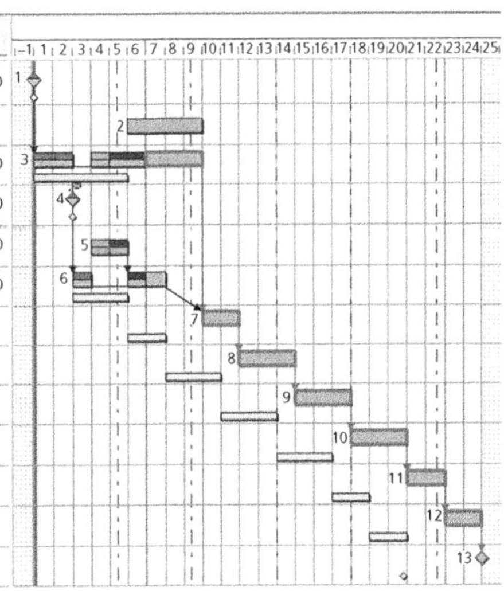

Owing to CE2, planned Completion current at the dividing date, taking into account any existing delay caused by the compensation event already in the Accepted Programme, moved to the end of week 24 and so the Completion Date will move to the end of week 24.

That is owing to the wording:

> The assessment takes into account.
>
> - any delay caused by the compensation event already in the Accepted Programme

So, the existing delay is recognised but there is no further delay to take account of. The existing delay was 4 weeks to planned Completion.

If, however, the effect of CE2 physical conditions had increased from that predicted at the end of week 4 and no other changes had removed CE2 from the critical path, then the effect of CE2 would be the delay already in the Accepted Programme and the further delay to the changed planned completion date as predicted at the dividing date.

> That would be captured by the wording
>
> - events which have happened between the date of the Accepted Programme and the dividing date.

7.10 Implications of the drafting from Example 5

From our more detailed analysis, we can see that the Accepted Programme current at the dividing date provides an agreed base line to start the process. However, we also have seen that at the dividing date that programme will be more or less up to date depending on intervals for submission and whether the *Project Manager* has accepted a revised programme or not responded.

There is a remedy if the *Project Manager* has simply not responded to a programme that has been submitted but that adds at minimum a further week to the point where we know the position, and longer if the *Contractor* is slow in reminding the *Project Manager* of the failure to respond.

At the point where the *Contractor* has to produce a programme for a quotation there are a number of additional considerations that have to be brought into the programme. It is here that the scope for disagreement can arise. For example, the assessment of a forecast or the extent to which delay in the programme has been caused by any one event as opposed to another event are matters of fact and judgement.

As we have seen in our example, the wording 'A delay to the Completion Date is assessed as the length of time that, due to the compensation event, planned Completion is later than planned Completion as shown on the Accepted Programme current at the dividing date' is deceptively simple.

It was criticised in NEC3 and has been updated in NEC4 and now we know that we add into that programme the further considerations discussed above where the *Contractor* is making its quotation.[63]

[63] See *Keating on NEC Contracts 2nd Edition* (Thomas and Lee, 2022: p. 443).

But it is in those steps that the areas of judgment are being exercised and also where we have to *assess the length of time that, due to the compensation event, planned Completion is later than planned Completion* (emphasis added).

In the next chapter, we will consider some of the other considerations when seeking to separate the effect of one compensation event from others and the implications of the contract drafting.

Willis T
ISBN 978-1-83662-481-3
https://doi.org/10.1108/978-1-83662-478-320251009

Chapter 8
The Wider implications of the NEC4 ECC

8.1. Introduction

In the last chapter, we saw some of the specific consequences of the drafting. As always with contracts, they exist in a legal context. In the context of the mechanisms to provide for extension of time, a great deal has been written by commentators about other standard forms of contract and the general approach of tribunals to this issue. Some of that commentary is based upon an approach to the question of extensions of time that, on the face of it, might seem at odds with the approach taken in the NEC4 ECC. The approach to what is referred to as concurrent delay is one area that might seem at odds and a possible resistance to the use of predictive analysis is another. We will consider this in this chapter.

8.2. How does the NEC4 ECC deal with concurrent delay?

Where one of the delaying events is not a compensation event and one is a compensation event then it becomes essential to assess 'the length of time that due to the compensation event planned Completion is later than planned Completion shown on the…'. (emphasis added)

This has some uncomfortable consequences for the *Contractor.*

Let us assume that the *Contractor* is in critical delay, so that the planned Completion date is put back meaning that the *Contractor* loses its terminal float.

Assume a compensation event such as additional work is instructed, which unless it is on the critical path would not allow any extension to the Completion Date as it had no effect on planned Completion.

What if our additional work does affect the critical path, but only so that it is equally a cause of planned Completion being late but not later than the planned Completion date owing to the *Contractor's* delay?

This question of whether when there are two delays both impacting Completion the contract the Completion date should be adjusted is often referred to as concurrent delay in the context of delay analysis.

One of the seminal documents in the context of the legal analysis of delay and related extensions of time is the Society of Construction Law Delay and Disruption Protocol 2nd Edition: February 2017. This is based upon the approach of the English courts to concurrency. It gives the following meaning of concurrency.[64]

'Meaning of concurrent delay

> *10.3 True concurrent delay is the occurrence of two or more delay events at the same time, one an Employer Risk Event, the other a Contractor Risk Event, and the effects of which are*

[64] Guidance Part B: Guidance on Core Principles in Society of Construction Law Delay and Disruption Protocol (SCL, 2017: p. 12).

felt at the same time. True concurrent delay will be a rare occurrence. A time when it can occur is at the commencement date (where for example, the Employer fails to give access to the site, but the Contractor has no resources mobilised to carry out any work), but it can arise at any time.

10.4 *In contrast, a more common usage of the term 'concurrent delay' concerns the situation where two or more delay events arise at different times, but the effects of them are felt at the same time.*

10.5 *In both cases, concurrent delay does not become an issue unless each of an Employer Risk Event and a Contractor Risk Event lead or will lead to Delay to Completion. Hence, for concurrent delay to exist, each of the Employer Risk Event and the Contractor Risk Event must be an effective cause of Delay to Completion (not merely incidental to the Delay to Completion).*

10.6 *This issue has both practical and legal implications. From a practical perspective, the analysis of the effects of the delay events is simpler if it considers only those events that will result in Delay to Completion (rather than a consideration of all events in the programme) so that the grant of an EOT follows the outcome of the critical path analysis. The Protocol recommends this approach during the currency of the project to allow the timely application for, and assessment of, EOT.'*

The first thing to note is that the references in the Protocol are to the Completion Date and not to planned Completion. This is important because it assumes that the entitlement to an extension of time only arises if there is a critical impact on the Completion Date. Consistent with that approach the treatment of terminal float is as follows.

'8. *Float as it relates to time Float values in a programme are an indication of the relative criticality of activities and, generally, when float is exhausted, the completion date will be impacted. <u>Unless there is express provision to the contrary in the contract,</u> where there is remaining total float in the programme at the time of an Employer Risk Event, an EOT should only be granted to the extent that the Employer Delay is predicted to reduce to below zero the total float on the critical path affected by the Employer Delay to Completion (i.e. if the Employer Delay is predicted to extend the critical path to completion).'*
(Underlines are author's own)

In the context of the NEC contract, we have seen that the contract does have express provision to the contrary in that our reference point is planned Completion, which preserves any terminal float.

However, the implications of the NEC drafting go beyond this. When we look at the analysis of concurrency and its relationship to extensions of time as propounded by the Protocol, we see that the analysis is very much influenced by the restraints that the law imposes upon the deduction of liquidated damages.

The Protocol sets out the two competing lines of argument derived from case law.

'10.7 *From a legal perspective, there are two competing views as to whether an Employer Delay is an effective cause of Delay to Completion where it occurs after the commencement of the Contractor Delay to Completion but continues in parallel with the Contractor Delay. This can be illustrated by the following example: a Contractor Risk Event will result in five weeks*

Contractor Delay to Completion, delaying the contract completion date from 21 January to 25 February. Independently and a few weeks later, a variation is instructed on behalf of the Employer which, in the absence of the preceding Contractor Delay to Completion, would result in Employer Delay to Completion from 1 February to 14 February.

10.8 On one view, the two events are both effective causes of Delay to Completion for the two week period from 1 to 14 February because they each would have caused Delay to Completion in the absence of the other (with the subsequent delay from 15 February to 25 February caused by the Contractor Risk Event alone). This view may be supported by older English appeal court cases (no doubt predating critical path analysis) which provide that if the failure to complete the works is due in part to the fault of both the Employer and the Contractor, liquidated damages will not be payable. In a situation like the example described in paragraph 10.7 above, it can be argued that both the SCL Delay and Disruption Protocol 2nd Edition: February 2017 31 Employer Risk Event and the Contractor Risk Event are in part the cause of the Delay to Completion.

10.9 On the other view, the Employer Delay will not result in the works being completed later than would otherwise have been the case because the works were already going to be delayed by a greater period because of the Contractor Delay to Completion. Thus, the only effective cause of the Delay to Completion is the Contractor Risk Event. This is the consistent position taken in recent lower level English court decisions.

10.10 The Protocol recommends the latter of these two views, i.e. that where an EOT application relating to the situation referred to in paragraph 10.7 above is being assessed, the Employer Risk Event should be seen as not causing Delay to Completion (and therefore there is no concurrency). Concurrent delay only arises where the Employer Risk Event is shown to have caused Delay to Completion or, in other words, caused critical delay (i.e. it is on the longest path) to completion. The Protocol cautions that this recommendation would have to be re-considered were an appeal court to take a different approach to this issue.

10.11 In considering whether concurrent delay exists, the Protocol recommends a common sense approach to delay analysis. In particular, the Protocol recognises that delay analysis is rarely precise down to the day (or even few days). The application of common sense requires that the margin for imprecision should be taken into account in reaching a conclusion on concurrency. Dealing with concurrent delay.

10.12 Where concurrent delay has been established, the Contractor should be entitled to an EOT for the Employer Delay to Completion, dealt with in accordance with Core Principle 5. The Contractor Delay should not reduce the amount of EOT due to the Contractor as a result of the Employer Delay.

10.13 An Employer should be aware that if it instructs a variation after the contract completion date where the failure to complete by the contract completion date has been caused by Contractor Delay, the Employer may lose its entitlement to liquidated damages if the Contractor then accelerates to recover the Contractor Delay to Completion at its own cost and that results in the variation (an Employer Risk Event) becoming the effective cause of Delay to Completion.

10.14 Employer Delay to Completion does not exonerate the Contractor for all its delays prior to that Employer Delay to Completion occurring. The effect of the Employer Delay should be assessed as described in Core Principle 5 and any EOT determined due should simply be added to the contract completion date.

> 10.15 *The Protocol's approach to dealing with concurrent delay aims to provide contracting parties with clarity and certainty about entitlement to EOT.*
>
> 10.16 *The Protocol's position on concurrent delay is influenced by the English law 'prevention principle', by virtue of which an Employer cannot take advantage of the non-fulfilment of a condition (for example, to complete the works by a certain date), the performance of which the Employer has hindered. The Protocol's approach to the treatment of concurrent delay (once established) SCL Delay and Disruption Protocol 2nd Edition: February 2017 32 prevents arguments about whether an Employer Delay acting concurrently with a Contractor Delay actually hinders the progress of the Contractor in any way.'*

As noted in the guidance to Core Principle 10 in Part B of the Protocol, where *Employer* Risk Events and *Contractor* Risk Events occur sequentially but have concurrent effects, the delay analysis should determine whether there is concurrent delay and, if so, that an EOT is due for the period of that concurrency. In this situation, SCL Delay and Disruption Protocol 2nd Edition: February 2017 26 states that any *Contractor* Delay should not reduce the amount of EOT due to the *Contractor* as a result of the *Employer* Delay. Analyses should be carried out for each event separately and strictly in the sequence in which they arose.

The concentration on the prevention principle and the ability of the *Employer* to deduct liquidated damages is a different focus to that of the NEC4 ECC. First, the date that the mechanism aims to preserve is the planned Completion date, which has no relevance to the deduction of liquidated damages – as opposed to the Completion Date, which is the reference point for liquidated damages.

Second, the extension of time drafting does not in all cases grant the *Contractor* an extension of time where there may be concurrent prevention occurring.

For example, where the *Project Manager* certifies take over under clause 35 because the *Client* has started using the works (and that was not envisaged in the Scope) before both Completion and the Completion Date, that is a compensation event. But note if the *Contractor* is in critical delay and the Completion Date has passed, there is no extension notwithstanding that the *Client's* actions might further increase the delay to the *Contractor* by disrupting its programme to completion.

The focus of the Protocol is very much upon interpreting the legal position to 'permit' an extension of time where there is a perceived risk that in the absence of such a power the mechanism could be considered to have broken down, such that the Completion Date became at large.

As the Protocol suggests, there are competing lines of authority as to how 'concurrency' is approached. In adopting the approach taken by lower courts and their guidance, the Protocol does take the more rigorous approach in that it suggests that the critical path analysis should determine whether there is true concurrency. Further that it will be rare. But the conclusion that the '*Contractor Delay should not reduce the amount of EOT due to the Contractor as a result of the Employer Delay*' would not be consistent with the contract drafting in an NEC ECC contract.

In the context of the NEC ECC the contractor has to show ***the length of time that, <u>due to the compensation event,</u> planned Completion is later than planned Completion.*** (emphasis added)

Planned Completion is a point in time and what has driven that is not relevant. Under our ordinary use of language if the compensation event does not move the planned Completion date to a later date there is no entitlement.

The right of parties to determine rules around concurrency is recognised by the law.[65]

The FIDIC contract invites as an optional statement for the parties to address concurrency and it is also dealt with in many bespoke contracts.[66]

Some commentators suggest that the NEC contracts along with JCT contracts do not address concurrency. But those statements appear to be based upon there being no specific reference to concurrency (unlike FIDIC, which simply leaves an option for it to be addressed).

However, if the basis upon which the extension of time can be granted excludes taking into account concurrent contractor/client delay, then the contract has dealt with concurrency but just not in the manner the commentators refer to.

8.3. Commentator's views of the concurrent delay issue and the use of predictive adjustment

The commentary on NEC3 in *Keating on NEC* (Thomas and Lee, 2022, 7-178) when considering the NEC 3 states as follows.

> *'As the contract provides a tool in the form of an up to date Accepted Programme for assessing delay, it is sometimes that suggested that delay must be assessed mechanistically. by inserting the individual compensation event into the Accepted Programme and reading off the Planned Completion. However, no one method of assessment is stipulated by the contract and the overarching requirement is probably that a reasonable and fair assessment is made. The Contractor and subsequently the Project Manager may decide that a mechanistic programme analysis as suggested above does not give a reasonable result and substitute such a result whilst continuing to take into account the programme analysis.'*

A link is then made in the footnote to the commentary on the assessment of Defined Cost and that the concept should be considered as in some way equivalent to damages and that would require a reasonable assessment to be made.

The commentary points out perceived difficulties with the drafting in NEC 3 that the cut off point for distinguishing between actual cost and forecast Defined Cost pre-dates the point in time at which the *Project Manager* considers the *Contractor's* quotation and suggests that nothing prevents consideration of later information available at the date of assessment.[67]

---

[65] North Midland Building Ltd v Cyden Homes Ltd [2018] EWCA Civ 1744.

[66] See clause 8.5 in *FIDIC Red Book* Conditions of Contract for Construction 2nd Edition 2017, SC 8.5, which allows the contract draughtsman to specify rules and procedures for concurrency in the Special Provisions and if not so stated 'as appropriate taking due regard of all relevant circumstances'.

[67] A link is then made to the commentary on the assessment of Defined Cost and that the concept should be considered as is some way equivalent to damages and that would require a reasonable assessment to be made.The commentary points out perceived difficulties with the drafting in NEC 3 where the cut off point for distinguishing between actual cost and forecast Defined Cost pre-dates the point in time at which the assessment is made and

Having made this link, the commentary continues.

> *'As set out above in respect of clause 63.1, the Project Manager appears to have a broad power under the second bullet point of clause 64.1 to decide that the Contractor has not assessed the compensation event correctly and is probably entitled to consider information then available in deciding whether the Contractor's assessment in his quotation is correct.'*

Note the language refers to deciding that a process has not been followed correctly. But should that decision be based upon information that post-dates the process?

What we know from the drafting is that the assessment is a prediction, and it is implicit that it may be 'wrong' or not correct with the benefit of hindsight or later information. But the test is whether the *Contractor* assessed the event correctly in the quotation.

The analysis has certainly 'watered down' the extent to which the drafting provides the basis for assessing the extension of time and then continues

> *'Clause 63.3*[68] *(the discussion is still about NEC3) does not provide any express guidance on how the issue of concurrent delay will be dealt with nor have there been any decided cases on the ECC wording and how concurrency is to be dealt with.'*

Reference is then made to the *Henry Boot case*,[69] which was a case based on the JCT 1980 Conditions.

The commentary continues

> *'The wording of clause 25 of the JCT 1980 is different from that of the ECC. Notably under clause 25.3.3 the Architect is entitled to reassess any extension of time within 12 weeks of practical completion. However, it is considered that what is common to both JCT clause 25 and ECC clause 63.3 is that the wording does not lay down hard and fast rules for the assessment of extensions of time. Therefore, it is understood that that the conclusion reached in Henry Boot is likely to apply to ECC and that the Project Manager and Contractor must take account of delays caused by other compensation events and delays for which the Contractor is responsible in assessing the delay reasonably attributable to each compensation event.'*[70]

Having analysed the drafting of NEC on the basis outlined the commentary then moves on to deal with concurrent delay and refers to the case law which is also considered in the SCL Delay and Disruption Protocol 2nd Edition (2017) and relies on cases concerned with a JCT standard form.

---

suggests that nothing prevents consideration of later information available at the date of assessment. It concludes that common sense also suggests that it would be wrong to ignore any actual costs that have been incurred or avoided by the time of the quotation.

68 63.3 A delay to the Completion Date is assessed as the length of time that, due to the compensation event, planned Completion is later than planned Completion as shown on the Accepted Programme. A delay to a Key Date is assessed as the length of time that, owing to the compensation event, the planned date when the Condition stated for a Key Date will be met is later than the date shown on the Accepted Programme.

69 Henry Boot Construction (UK) Ltd v Malmaison (Manchester Limited) (1999) 70 Con L.R.32 (TCC).

70 7-178 433 and 444.

The commentary suggest that concurrency must be dealt with in a '*fair and reasonable way*' by the '*decision maker*' and also notes the slight difference of approach between the English and Scottish courts to concurrency where two causes are operative, one being a relevant event (the JCT term of events entitling the Contractor to an extension of time) and some other event for which the Contractor is to be taken as responsible and neither of which could be described as the dominant cause. In the Scottish case of *City Inn*,[71] the court stated the claim for extension of time will not necessarily fail and it will be open for the decision maker, be that the architect or other tribunal approaching the issue in a fair and reasonable way, to apportion the delay to the completion of the works between the relevant event and the other event. The court noted *'it must be recognised that the background to the decision making, in particular the possibility of a claim for liquidated damages, as opposed to one of extension of time must be borne in mind and approached in a fair and reasonable manner'.*

The English law approach to this situation is that where two causes have equal potency, and each party is responsible for one, the contractor would be an entitled to an extension of time.

So, it is clear from the commentary that the authors consider that the general rules relating to concurrent delay applied to the NEC3 ECC.

There have, however, been some significant changes to the drafting in the NEC4 form of contract and the suggested approach to the NEC3 ECC contract may well not be applicable. The equivalent clause to ECC3 63.3 is now ECC clause 63.5.

> *'63.5 A delay to the Completion Date is assessed as the length of time that, due to the compensation event, planned Completion is later than planned Completion as shown on the Accepted Programme current at the dividing date. A delay to a Key Date is assessed as the length of time that, due to the compensation event, the planned date when the Condition stated for a Key Date will be met is later than the date shown on the Accepted Programme current at the dividing date.'*[72]

> The assessment takes into account
>
> - any delay caused by the compensation event already in the Accepted Programme and
> - events which have happened between the date of the Accepted Programme and the dividing date.

The commentary on NEC4 states '*The amended wording to ECC4 clause 63.5 goes some way to addressing some of the issues arising from the wording ECC3 clause 63.3.*'

The commentary points out that the base line programme is now the 'Accepted Programme current at the dividing date'. The commentary recognises that under NEC4 there will usually be an up-to-date Accepted Programme because of the new default mechanism that applies if the *Project Manager* has not responded to a programme submitted. The commentary continues.

---

[71] City Inn Ltd v Shepherd Construction Ltd [2010] B.L.R. 473.

[72] The reference to the Accepted Programme current at the dividing date is necessary owing to the definition of the Accepted Programme. The definition of the Accepted Programme is 11.2(1) 'the programme identified in the Contract Data or is the latest programme accepted by the *Project Manager*. The latest programme accepted by the *Project Manager* supersedes the previous Accepted Programmes'.

> *'Furthermore, the Contractor and the Project Manager is not simply required to mechanistically add in delay caused by the compensation event without regard for other matters.'*

By reason of new clause 63.5, the first bullet point, the *Contractor/Project Manager* must take account of any delay already caused by the compensation event already in the Accepted Programme.

The second bullet point expressly states that the *Contractor/Project Manager 'takes account of events which have happened between the date of the Accepted Programme and the dividing date'.* The commentators also refer to actual delay and predicted delay caused by a previous compensation event and conclude that such delays would be captured in the requirement for revised programmes to show actual progress and planned Completion and so it would be inconsistent if that was not also something to be taken into account in the second bullet point.

That is clearly correct in that the programme should always reflect the facts, the 'as-built' position and the predicted planned Completion date. Whether an operation is part of work impacted by a compensation event or some other delay the programme simply shows where that operation has got to and how any impacts will affect planned Completion.

If a preceding compensation event assessed the impact of an event as effecting certain operations to give new finish dates for those operations and they are taking longer than originally assessed, the programme – whether revised under clause 32 or updated under 62.2 – should show their true finish dates. If those are later than had been planned then planned Completion will be later (assuming they are on the critical path), but there will be no adjustment to the Completion Date as the delay is not owing to a new compensation event and the assessment of the compensation that was the source of the delay has been assessed already. It is not revised because the forecast was wrong.

The commentators say nothing specific about the analysis of 'concurrent delay' in NEC4 ECC. Given that the arguments the commentators raised in support of a traditional approach to concurrency in the NEC3 ECC were based largely on the conclusion that the contract failed to deal with the method of assessment – stating: *'However no one method of assessment is stipulated by the contract and the overarching requirement is probably that a reasonable and fair assessment is made'* and that their analysis now is that the amended wording *'goes some way to addressing some of the issues arising'* – would the suggestion that the contract does not stipulate a method of assessment hold?

As we have seen, there is no objection in principle to the contract excluding concurrent delay as a basis for an extension of time and the wording of clause 63.5 points to the compensation event having the effect that planned Completion is later than planned Completion as analysed in the programme supporting the extension of time.

The compensation event must move the planned Completion date. On our analysis of concurrent delay being equally operative causes, then the compensation event has not moved the planned Completion date if a *Contractor* delay has moved the date. If they both move the date, and are absolutely equally operative in time, then no extension would be given unless we read into the contract a rule of law that overrides the drafting.

8.4. Concurrent delay scenario example

For example, if we have a base line programme, as illustrated below, submitted at the tender stage. See Figure 8.1.

Assume we are now week 4 of the contract and Activity A is 3 weeks behind programme. The revised programme submitted at the end of week 4 shows the actual progress. The effect of the delay has been to use up the terminal float of 1 week and planned Completion has gone back from beginning of week 15 to beginning of week 17. See Figure 8.2.

Figure 8.1 Base line programme submitted at the tender stage (© Copyright nec 2025)

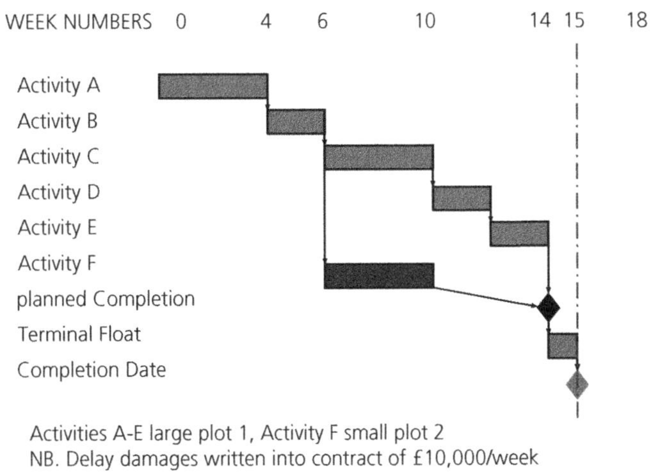

Figure 8.2 Revised programme submitted at the end of week 4 (© Copyright nec 2025)

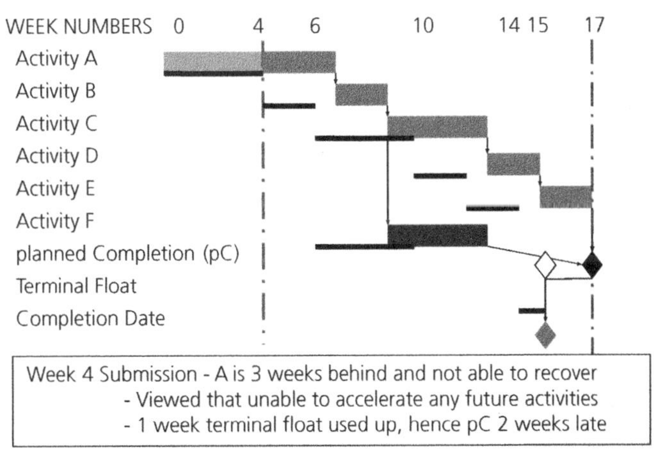

The programme is submitted and in week 5 is accepted and becomes the Accepted Programme. See Figure 8.3.

In week 6, the *Project Manager* issues a compensation event which extends Activity F by 3 weeks. The *Contractor* has to submit a quotation and should show any effect on the remaining work.

The dividing date is the date of the instruction, which is week 6. The baseline programme for assessing the compensation event is the Accepted Programme current at the dividing date. See Figure 8.3. When we impact the compensation event it has the effect that Activity F now moves onto the critical path. But this has no effect on planned Completion, which remains at week 17. See Figure 8.4.

Figure 8.3 Revised programme accepted and becomes the Accepted Programme (© Copyright nec 2025)

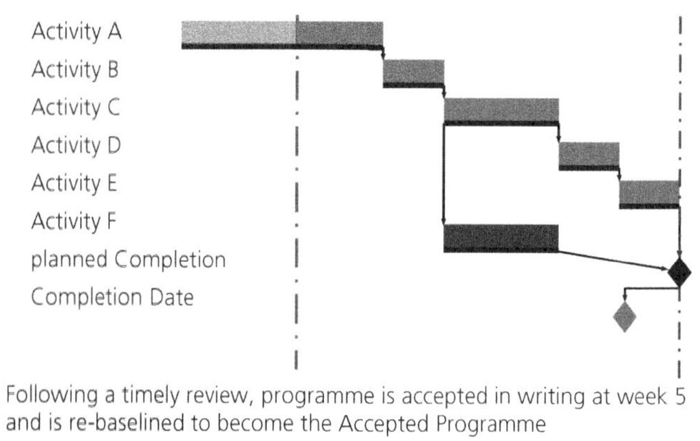

Figure 8.4 Impact of the compensation event (© Copyright nec 2025)

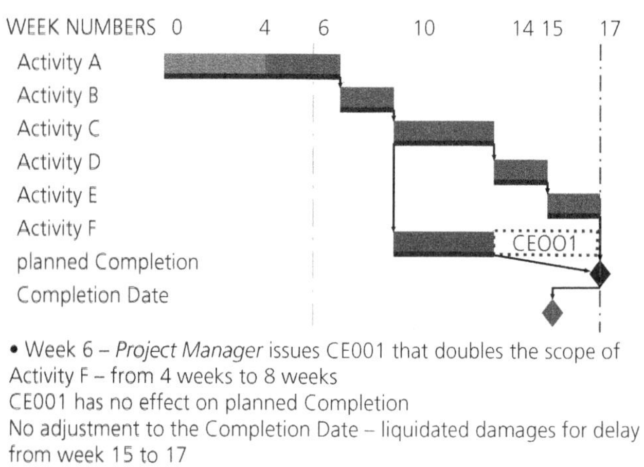

Assuming the contract had delay damages and based on the analysis that any extension of time the *Contractor* is entitled to has to be based on the contract wording

'*A delay to the Completion Date is assessed as the length of time that, due to the compensation event, planned Completion is later than planned Completion as shown on the Accepted Programme current at the dividing date.*'

The implications would seem to be that under the drafting, the position would appear to be that there is no entitlement to an extension and the delay damages would be recovered by the *Client*.

If the *Contractor* could accelerate the delayed activity then it might be able to suggest in its 'updated' programme that its planned Completion for its activities, apart from the compensation event, is now the Completion Date; so, the compensation event is now the critical delay and, therefore, it is entitled to an extension.

But if the *Contractor* cannot do that then liquidated damages would be taken.

On the basis of a traditional analysis, the *Contractor* would argue that the additional work controls planned Completion as much as the pre-existing delay. They are equal operative delays and it is, therefore, entitled to more time.

8.5. Does the NEC4 ECC recognise concurrent delay? Discussion

We saw that the commentary in *Keating on NEC*[73] considers that NEC4 has given further content to the mechanism for adjusting the Completion Date and recognises the dividing date as being more specific in determining which programme we use as our base line.

The dividing date is the logical date to choose for a contemporaneous and prospective delay analysis and gives the rules for the *Contractor* to make its quotation. It is on that basis that the *Project Manager* is expected to assess whether the *Contractor* has made its assessment correctly. The suggestion raised in the NEC 3 ECC commentary that the *Project Manager* at the point where they consider that quotation can consider actual delay after the dividing date is, in reality, taking us into a different process where the *Project Manager* is 'assessing' the delay. The point is that the *Project Manager* only makes their own assessment if certain conditions have been met. There is nothing in the drafting to support the view taken in the NEC3 ECC commentary by the commentators that the *Project Manager* must consider all information at the point where the *Project Manager* is considering the *Contractor's* quotation.

The dividing date forms a cut-off point and division between forecasts for which the *Contractor* is taking the risk and actual delay. If the risk is overstated in principle, then it is an incorrect assessment which the *Project Manger* would not accept. But if it is just shown by later records to be an underestimate or overestimate, the *Project Manager* should not go behind the drafting and treat it as an incorrect assessment. The *Project Manager* should also be consistent. If the actual records show the risk allowed in the forecast was too small that would not be a basis for not accepting the quotation.

[73] *Keating on NEC, Second edition* (Thomas and Lee, 2022: pp. 442–444; 445–446)

These are, in a sense, fine distinctions but what the contract envisages as the starting point is a *Contractor*'s assessment in the form of the quotation with rules as to how that quotation is built up.

There has to be a cut-off point for the actual information it can include. It cannot be the date at which the *Project Manager* is considering the quotation for acceptance or not because that becomes, in a sense, a *Project Manager's* assessment (which is a different process under the contract). The quotation if accepted is meant to bind the *Contractor*.

The suggestion, made in the commentary on NEC3, that the *Project Manager* should ignore the dividing line and consider information about 'actual' delay after the dividing date and on that basis would be entitled to not accept the quotation, is not made in the commentary on NEC4.

The commentary on NEC 4 also leaves open the question of the treatment of concurrent delay in the NEC4 ECC.

Unless we can say that the mechanism is problematic, as was suggested to be the case for the NEC3 ECC, then there would appear to be no reason to ignore the contract drafting and apply some other rule.

If we do not ignore the drafting and supplant it with some other rule then, for the reasons discussed, it is considered that the *Project Manager* when accepting or not accepting the *Contractor's* quotation should not grant any adjustment to the completion date where a *Contractor* delay and a *Client* delay are of equal operative effect, so that both control the planned Completion date on the analysis.

A slightly different but related point is the requirement generally to act without fear or favour and so to follow the analysis and not be driven by the outcome.

As we shall see, when we come to the dispute resolution provisions it may be arguable that, depending on the nature of the dispute process, different rules would apply where someone other than the *Project Manager* has to make the adjustment to the Completion Date in a dispute about an extension of time.

Willis T
ISBN 978-1-83662-481-3
https://doi.org/10.1108/978-1-83662-478-320251010

Chapter 9
Keeping the Accepted Programme

9.1. Introduction

The Accepted Programme is intended to sit at the heart of the NEC risk management process and to provide an agreed basis for adjusting the Completion Date. It is apparent that the process of updating and agreeing the revised programmes can itself be challenging. In this chapter, we consider how to make the process more likely to succeed.

9.2. Best practice in keeping the Accepted Programme as a collaborative risk management tool

The contract drafting allows for the programme to be revised and sets out a minimum requirement that it is revised at the interval stated by the *Client* in the Contract Data Part One. The programme can also be revised whenever the *Project Manager* instructs, or the *Contractor* chooses. Where the *Project Manager* instructs, the revised programme must be submitted within the *period for reply*. The submission procedure envisages that a programme is reviewed within 2 weeks of submission. These standard provisions set out minimum requirements but it is clear that if all that is done is that the *Contractor* complies with this and if, as on many projects, the interval set for updating the programme is 1 month, the programme can be considerably out of date at the point where an assessment is needed or the impact of delay is being considered in an early warning meeting.

Therefore, timescales should be kept to a minimum necessary to keep the programme up to date. The timescales can be improved upon if the *Project Manager* is able to quickly assimilate any changes to the programme and if the information provided to support the programme is transparent and reliable.

The drafting of the standard form recognises that the requirements in the contract may need to be supplemented. Clause 31.2 requires that the *Contractor* shows on each programme submitted for acceptance 'other information which the Scope requires the *Contractor* to show on a programme submitted for acceptance'.

The guidance Volume 2 at page 66 suggests that the programme requirement are considered and addressed in the Scope.

S 500 Programme

> The ECC includes detailed programme requirements and procedures. It may be necessary to set out specific *Client* requirements.

Checklist	Explanation
S 505 Programme requirements **ECC 31.2**	State requirements for the format of the programme, including the use of specific software (if necessary) and the requirement for hard or electronic copies. State any requirement for the programme to be produced in levels (for example summary level to detail level).
ECC 31.3	State any information additional to the requirements of clause 31.2 that the *Contractor* is to show on the programme. This may include dates for submission of designs and samples, dates for information or actions by the *Client* and *Project Manager*, and the timing of any test and inspection.
S 510 Methodology statement	Particular requirements for methodology statements, including any specific requirements for the format of resource information.
S 515 Work of the *Client* and Others **ECC 25.1** **ECC 60.1(5)**	Detail the order and timing of the work of the *Client* and *Others* to be included in the programme and information to be provided. Refer as necessary to sections S 905 and S 910.
S 520 Information required	A schedule of information to be provided, who it is to be provided by and the date by which it is to be provided.
S 525 Revised Programme	State any specific requirements for the submission of revised programmes such as an explanation of changes.

The Scope could prescribe the form of the programme to be submitted and the access that is to be provided to the software to the *Project Manager*.

The suggested additional information should be easily available to the *Contractor* and should include

> Method statements explaining the use of the equipment and resources and identifying the equipment and resources to be used.

Some commentators have suggested that in a UK context the method statements that are prepared in the context of the CDM regulations will provide the basis for understanding the resourcing assumptions. That might be the case depending on the nature of the project, but on some projects there will not be a direct read across from that information to all the operations in the programme. Reliance on the information provided for another purpose may not be enough.

In addition, for revised programmes the information would ideally include

- Supporting progress records to demonstrate actual progress.
- Any area of delay to be identified and a statement included as to how the *Contractor* plans to deal with it.
- Notified defects to be identified and the plans for correction of those included.
- The reasons for any other changes to the programme to be provided including any resourcing or linkage information.

It is suggested that if these measures are taken then there would be little reason for there to be any disagreement over the programme and that a programme could be very quickly accepted.

9.3. Co-hosting/Collaborative programme management

The intention behind such measures is to seek to provide the background against which agreement of the programme is likely to be most effectively achieved.

These requirements highlight the extent of programming that the *Contractor* is required to do and also the resources that both the *Client* (via the appointed *Project Manager*) and the *Contractor* need to commit to the project. Some project managers and some contractors suggest that this degree of resourcing is not realistic. This fails to recognise that if the contract is managed as the drafters intend then that adherence or non-adherence to the process is not a choice for them to make.

A client must make that choice when is selects the NEC4 ECC for delivery of its project.

If the importance of the programme is recognised then the *Client* should include detailed requirements in the Scope, addressing the programme form and content and the related issue of progress records.[74]

In addition to the sections of the Scope dealing with the programme, the NEC4 ECC and other contracts provide for the use of a communications system.

See clause 13.2 and section S800 of the Guidance Volume 2 example Scope. These sections of the Scope can more fully address requirements for co-hosting systems and co-location.

The guidance states the following.

> The ECC establishes a procedural framework based on good project management practice.
>
> It may be helpful to detail the management and communication procedures required to support this. This may include a framework of regular meetings, attendees required and outputs. Explain how people will be involved in the management of the works and how communications are to be managed. Consider the use of a chart setting out the roles and responsibilities of the various parties involved.

Further sections deal with other matters.

S 810 and 815 deal with Communication System and Management Procedures and states the following.

[74] There are a number of examples of suggested progress records such as: Society of Construction Law Delay and Disruption Protocol 2nd Edition: February 2017 (SCL, 2017); CIOB (2018) *Guide to Good Practice in the Management of Time in Major Projects* (CIOB, 2018).

Detail the communication system to be used. Consider the use of the following:

- Internet based collaboration tool.
- Electronic mail system or
- Standard forms and templates.

S 815 Management procedures

State any management procedures which the Contractor is required to follow. Consider the following:

- Meetings, attendees and meeting records.
- Reporting requirements (e.g. progress reports).
- Information requirements.
- Terminology and abbreviations.

If such requirements are considered unnecessary, the NEC Short Contract provides the basis for delivery of projects that do not have complex programming or risk management needs. There are also other standard forms outside of the NEC family where the programme has little or no relevance to the day-to-day administration of the contract.

NEC4: Effective Preparation and Management of the Construction Programme

Willis T
ISBN 978-1-83662-481-3
https://doi.org/10.1108/978-1-83662-478-320251011

Chapter 10
Project Managers' assessments of adjustment to the Completion Date

10.1. Introduction

The NEC4 ECC seeks to provide a mechanism to reach an agreed position on the programme and any adjustment to the completion date. That will not always be possible. In this chapter, we consider the default position where the *Project Manager* must make their own assessment.

10.2. Best practice for *Project Managers* when making their own assessment

Although the contract drafting envisages that the outcome of the compensation event procedure will be an agreed quotation that deals with both the time and money implications of a compensation event, there are numerous circumstances where that will not be the case.

In these circumstances, there are generally two possibilities: either that the *Project Manager* will make their own assessment or the contract conditions (following a warning from the *Contractor*) deems something to have occurred.

The deemed occurrences will have arisen by virtue of the *Project Manager* having failed to do something within the requisite timescale and then failing to remedy that when notified within a further timescale. These are important provisions that attempt at least in part to preserve the contract procedures when the *Project Manager* is failing to do their part. We will consider the implications of that in Chapter 11 concerning disputes and whether the *Client* is bound by such deeming provisions.

For present purposes, we will concentrate on circumstances where the *Contractor* has failed to do something in which case it is the *Project Manager* who must take charge.

10.2.1 When does the Contract require a *Project Manager's* assessment of an extension of time?

The contract drafting refers to these circumstances both in relation to the process as it concerns Defined Cost and the assessment of any adjustment to the Completion Date or the Key Date.

Clauses 64.1 and 64.2 address this issue and consider failures of the 'agreement' process in terms of both the assessment of time and money.

However, the two issues are related and, as we shall see, the drafting does not suggest that a *Project Manager* could accept a quotation to assess financial adjustment but make their own assessment of time.

Clause 64 deals with assessing the compensation event and provides the following.

> 64.1 The *Project Manager* assesses a compensation event
>
> - if the *Contractor* has not submitted the quotation and details of its assessment within the time allowed,
> - if the *Project Manager* decides that the *Contractor* has not assessed the compensation event correctly in the quotation and has not instructed the *Contractor* to submit a revised quotation,
> - if, when the *Contractor* submits quotations for the compensation event, it has not submitted a programme or alterations to a programme which the contract requires it to submit or
> - if, when the *Contractor* submits quotations for the compensation event, the *Project Manager* has not accepted the *Contractor's* latest programme for one of the reasons stated in the contract.

As we can see, clause 64.1 concentrates on problems with the quotation process.

The first bullet point covers the position where no quotation has been submitted in time.

The second bullet point deals with the quotation not having correctly assessed the compensation event.

The third bullet point refers to clause 62.2 that a programme should be submitted to support the delay impact analysis if the programme for the remaining work is altered by the compensation event.

The fourth bullet point is a reference to the Contractor's latest clause 32 programme that has not been accepted for one of the reasons in the contract.

The related clause 64.2 concentrates on the question of adjusting the completion date based on the assumption that there is no agreed basis to start from in terms of programme and provides the following.

> 64.2 The *Project Manager* assesses the programme for the remaining work and uses it in the assessment of a compensation event if:
>
> - there is no Accepted Programme,
> - the *Contractor* has not submitted a programme or alterations to a programme for acceptance as required by the contract or
> - the *Project Manager* has not accepted the *Contractor's* latest programme for one of the reasons stated in the contract.

As can be seen from the drafting, there is an overlap between the status of the programme in clause 64.2 and the circumstances where the *Project Manager* will have to make their own assessment in clause 64.1.

The first bullet point in 64.2 is in addition to the drafting in clause 64.1. This covers the circumstance where there is no Accepted Programme. This would cover the position where a programme had not been submitted at all or no programme has ever been accepted.

The wording of the second and third bullet points is very similar to bullet points 3 and 4 in clause 64.1 but omits the wording 'if, when the Contractor submits quotations for the compensation event.'

In clause 63.5, where the contract refers to the base line for the analysis as being the Accepted Programme current at the dividing date, that wording could be broad enough to suggest that any previous Accepted Programme could form the base line. On that analysis, a contract could be in month 4 of the contract where the only Accepted Programme was the one accepted at tender and subsequent revisions had been rejected by the *Project Manager* leaving the tender programme as the Accepted Programme current at the dividing date. By introducing a *Project Manager's* assessment where a revision has not been accepted for the reasons stated in the contract, or a revision has not been submitted at all, means that the assessment based upon the '*Accepted Programme current at the dividing date*' would only occur where that programme is part of an agreed line of programmes as opposed to an isolated an out-of-date document.

The guidance note suggests that the fact that a *Project Manager's* assessment will follow from a failure to make compliant revisions is an added incentive for the *Contractor* to comply with the programming requirements.

Also, in the context of assessing compensation events, a programme is to be submitted under clause 62.2, which is an update on the Accepted Programme Current at the dividing date. If we read bullet points 2 and 3 as only referring to the clause 32 process, we do not have any reference to the clause 62.2 process. The bullet points are wide enough to cover both scenarios. Further, clause 64.1 would apply if there is no programme submitted as part of the quotation.

The aim of the drafting is to identify the circumstances in which the *Project Manager* either does not have a base line or the clause 62.2 updating process has failed so that the assessment of the programme implications falls to be made by the *Project Manager*.

The wording simply states 'The *Project Manager* assesses the programme for the remaining work and uses it in the assessment of a compensation event if.'

There is no further guidance and the 'programme for the remaining work' is used only to assess delay and the adjustment to the Completion Date for compensation events. It does not become the Accepted Programme.

The implementation of the compensation event does not itself involve the acceptance of the programme by either party. Clause 66 provides that a compensation event is implemented when

- 'The *Project Manager* notifies acceptance of the *Contractor's* quotation
- The *Project Manager* notifies the *Contractor* of an assessment made by the *Project Manager* or
- A *Contractor's* quotation is treated as having been accepted by the *Project Manager*.'

Clause 66.2 provides 'When a compensation event is implemented the Prices, the Completion Date and the Key Date are changed accordingly.'

Again, no mention is made of the process creating a new programme or an Accepted Programme.

As noted in the practice note, the clause 62.2 programme may not cover all the content that a clause 32 programme requires. In that case, the *Contractor* may have to submit a clause 32 programme for Acceptance. The *Contractor* can do so whenever it chooses (see 32.2) or the *Project Manager* may instruct that a revision is produced.

However, that would be a revision to the Accepted Programme, which might be a different document to the Accepted Programme current at the dividing date.

So, if we do not have an Accepted Programme current at the dividing date or a clause 62.2 update to it, on what basis does the *Project Manager* assess the compensation event?

10.2.2 The basis of the *Project Manager's* assessment

The concept of planned completion is now dropped in the drafting and the reference is *'the programme for the remaining work.'*

> 'The *Project Manager* assesses the programme for the remaining work and uses it in the assessment of a compensation event.'

The contract drafting in this regard is very limited. We no longer have a method for ascertaining the extension. The guidance note points out that this is not where either party would wish to be. The *Contractor* has lost 'the Accepted Programme' and the *Project Manager* must take on more responsibility, keenly aware that any ascertainment made must be capable of surviving challenge in adjudication.

Are there any rules that apply?

First, the *Project Manager* must be able to show that they have followed the process. This will require a record of their programme for the remaining work. The drafting does not suggest that any particular method is used or that any particular materials are used.

There are differences to the approach that the drafting requires is followed for *Contractor's* quotations, which raise questions to which there may not yet be answer.

Does the end date of the remaining work simply become a substitute for 'planned Completion'?.

If that is earlier than the Completion Date, is the *Contractor* entitled to the benefit of the terminal float?

Is the assessment made by the *Project Manager* a forecast in the sense that the *Contractor's* assessment must be and must allow for risk and if so must the *Project Manager* also allow for risk?

What if the effects of the compensation event are too uncertain to be reasonably forecast? Should the *Project Manager's* assessment be based on stated assumptions that can be reviewed if they later turn out to be wrong, as in clause 61.6?[75]

[75] There is some support for these rules still applying to be found in *Keating on NEC 2nd Edition* (Thomas and Lee, 2022). See 7-256, p. 463. But the wording of clause 61.7 does not fit with this. It assumes the *Project Manager* has issued an instruction.

The timescale for making an assessment is set (3 weeks starting from the time the need for the assessment becomes apparent), and so without the benefit of clause 61.6 the *Project Manager* may feel very exposed in trying to include risk in an assessment.

That is compounded by the fact that clause 66.3 states that 'The assessment of an implemented compensation event is not revised except as stated in these conditions of contract.'[76]

The *conditions of contract* do permit an assessment to be challenged and the wider implications if there is a dispute are considered in the next chapter.

The guidance simply states that the *Project Manager* will be motivated to make a fair and reasonable assessment in the knowledge that the *Contractor* may formally dispute the matter.[77]

From the limited drafting, it is clear that the *Project Manager* will have to some extent to fill in the gaps.

[76] *Keating on NEC 2nd Edition* (Thomas and Lee, 2022: p. 469).

[77] NEC4 Managing an Engineering Construction Contract Volume 4: p. 66.

Willis T
ISBN 978-1-83662-481-3
https://doi.org/10.1108/978-1-83662-478-320251013

Chapter 11
Implications of losing the Accepted Programme

11.1. Introduction

The loss of the Accepted Programme means that the parties are no longer in agreement about some important aspect of the project. It could be that the critical path is not agreed or that the dates by which certain activities were completed are not agreed or the disagreement could be about the future progress of work. The extent to which the parties have divergent views of the programme will vary. From both parties' perspective, getting to an agreed position so there is an Accepted Programme in place should be a priority. Failure to agree a programme will hamper the ability to agree how risk will impact progress and deprive the parties of a valuable tool for planning the response to risks. In this chapter, we look at the implications for the parties of losing the Accepted Programme.

11.2. Loss of the Accepted Programme adjustment to the Completion Date and other implications for the *Contractor*

From the *Contractor's* perspective, lack of an Accepted Programme will mean that future assessments of compensation events will be made by the *Project Manager* and that any challenge to those will have to be made through adjudication. This impacts not just extensions of time but also assessment of the financial compensation. See clauses 64.1 and 64.2.

In short, the *Contractor* must live with the *Project Manager's* assessment or refer a dispute to adjudication. There may be further commercial implications for, as we have seen, the basis of entitlement to both time and money changes.

In addition, several compensation events use the Accepted programme as their reference point. The loss of those reference points potentially moves the risk allocation for those events to the *Contractor*. Further, 'planned completion' may no longer have any significance in protecting terminal float.

If the *Contractor* has to challenge the *Project Manager's* assessment and the analysis of delay becomes 'retrospective', the basis to entitlement may become less certain as the area of dispute could become very broad.

The only potential commercial advantage to the *Contractor* might be that if the prospective method of assessing an extension of time does not allow an extension for concurrent delay, then a challenge to a *Project Manager's* assessment might hold out more chance of such an extension. But we only need to look at what we mean by concurrent delay and the acknowledged position that it is very rare to see that this is not such a significant benefit to the *Contractor* to 'game' the contract and submit a deliberately incorrect quotation or stop revising its programme.

For the *Client,* the loss of the Accepted Programme raises uncertainty in terms of the how robust the Completion Date and price for the work is at any point. Although the *Project Manager's* assessment will give a contractual position, that is not an agreed position, and it could be challenged through adjudication. As we have seen from the Chapter 10 Section 10.2.2, the basis upon which a *Project Manager's* assessment is made is not as clear as the rules governing the assessment of a quotation.

Where the *Contractor* has warned the *Project Manager* and a programme or compensation event has been deemed to be accepted, then it appears the *Client* may challenge that (see Chapter 10) but the dispute process is not something that any *Client* would wish to have to initiate.

11.3. Loss of the Accepted Programme implications for disputes over adjustments to the Completion Date and wider implications

11.3.1 The importance of the dispute resolution clause

If the Parties no longer have an agreed position, then it is possible that either party may wish to challenge the contractual position as determined by the contract. What that will entail and how such matters will be dealt with depends in part on the choice of dispute resolution clause. The clauses in question are W1 and W2.

11.3.2 W1

It is worth remembering that clause W2 is the 'newcomer' introduced into the contract in order to comply with the requirements for the statutory right to adjudicate, introduced in the Housing Grants Construction and Regeneration Act 1996. For most work procured under NEC ECC contracts in England and Wales and Scotland the relevant clause will be W2. But it is important to look at clause W1 to understand how the contract works without the statutory intervention.[78]

W1 is the clause that was originally included in the contract as a contractual dispute resolution process, and it has certain features that are intended to align the dispute resolution process and the broader aims of the contract. This can best be seen in the way the clause is set out. It uses a table to set out the types of dispute that might be referred and timescales for referring them.

(4) No evidence of the statement of case or discussions is disclosed, used or referred to in any subsequent proceedings before the *Adjudicator* or the *tribunal*.

DISPUTE REFERENCE TABLE

DISPUTE ABOUT	WHICH PARTY MAY REFER IT TO THE *SENIOR REPRESENTATIVES*?	WHEN MAY IT BE REFERRED TO THE *SENIOR REPRESENTATIVES*?
An action or inaction of the *Project Manager* or the *Supervisor*	Either Party	Not more than four weeks after the Party becomes aware of the action or inaction
A programme, compensation event or quotation for a compensation event which is treated as having been accepted	The *Client*	Not more than four weeks after it was treated as accepted

[78] The Housing Grants Construction and Regeneration Act 1996 as enacted and as amended cannot be contracted out of.

(Continued)

DISPUTE REFERENCE TABLE		
DISPUTE ABOUT	WHICH PARTY MAY REFER IT TO THE *SENIOR REPRESENTATIVES*?	WHEN MAY IT BE REFERRED TO THE *SENIOR REPRESENTATIVES*?
An assessment of Defined Cost which is treated as correct	Either Party	Not more than four weeks after the assessment was treated as correct
Any other matter	Either Party	When the dispute arises

The effect of the drafting is to require that disputes are referred in a relatively short timescale from the point that a party is aware of something it wishes to dispute, or something has been treated as accepted. Failure to comply with the timescales bars a Party from raising the disputed matter. In the context of the value of compensation events or the Completion Date being challenged, these would fall under the first category in the table and the timescale is not more than 4 weeks after the Party became aware of the event.

In that context, a dispute about a compensation event would be referred very quickly after the *Project Manager* had made their assessment or potentially be barred.

The first step is to notify of the dispute. Given the timescale of 4 weeks from becoming aware of the matter for giving notice for most matters concerning time related disputes and the timetable for then dealing with the dispute, the *Senior Representatives* will be dealing with up-to-date information when seeking to agree matters. See clauses W1.1(2) and W1.1(3).

> 'The Party referring a dispute notifies the *Senior Representatives*, the other Party and the *Project Manager* of the nature of the dispute it wishes to resolve. Each Party submits to the other their statement of case within one week of the notification. Each statement of case is limited to no more than ten sides of A4 paper together with supporting evidence, unless otherwise agreed by the Parties.
>
> The *Senior Representatives* attend as many meetings and use any procedure they consider necessary to try to resolve the dispute over a period of no more than three weeks. At the end of this period the *Senior Representatives* produce a list of the issues agreed and issues not agreed. The *Project Manager* and the *Contractor* put into effect the issues agreed.'

This gives an 8-week timetable from becoming aware of the matter to the point where the *Senior Representatives* have concluded their efforts at resolution.

If a Party still requires a matter that has not been agreed to be decided, then it must refer the matter to the *Adjudicator*.

> 'W1 3 (1) A Party disputing any issue not agreed by the *Senior Representatives* issues a notice of adjudication to the other Party and the *Project Manager within two weeks of the production of the list of agreed and not agreed issues, or when it should have been produced.* The dispute is referred to the *Adjudicator* within one week of the notice of adjudication.'

Against that timetable (11 weeks at most before the reference to the *Adjudicator*), the *Adjudicator* should be considering the case within 3 months of the dispute arising.

In terms of the procedure in the adjudication the drafting also points to specific powers[79]

> 'WI.3 (5) The *Adjudicator* may:
>
> - review and revise any action or inaction of the *Project Manager* or *Supervisor* related to the dispute and alter a matter which has been treated as accepted or correct,
> - take the initiative in ascertaining the facts and the law related to the dispute,
> - instruct a Party to provide further information related to the dispute within a stated time and
> - instruct a Party to take any other action which is considered necessary for the *Adjudicator* to reach a decision and to do so within a stated time.'

Specifically with regard to compensation events, W1.3 (7) states

> 'If the *Adjudicator's* decision includes assessment of additional cost or delay caused to the *Contractor*, the assessment is made in the same way as a compensation event is assessed.'

In a sense, the drafting is assuming that the *Adjudicator* will step in to do what the *Project Manger* should have done and failed to do. The Adjudication is more akin to a re-run of the compensation event process with the *Adjudicator* replacing the *Project Manger*.

This approach to the drafting can be seen in the Judgement of the Honourable Mr Justice Ramsey in *WSP CEL LIMITED v DALKIA UTILITIES SERVICES PLC* [2012] EWHC 2428 (TCC) at paragraphs 85 and 86.

> '85 *Mr Soole QC submitted that claims which were made by WSP for any matters which are compensation events have to be dealt with under Grounds 1 and 2 of the Adjudication Table as these relate to an action of the Employer or the Employer not having taken an action, generally at the end of the compensation event machinery. He submitted that therefore the dispute as to validity and value of the claim in the Final Account would all have to be dealt with under Grounds 1 and 2 and that the provisions of the NEC Conditions do not contemplate there being a large number of disputes which only arise at a later date, such as the date of termination and give rise to a dispute on a composite claim or on a final account at termination.*
>
> 86 *I consider Mr Soole QC's submission to be correct. The philosophy of the NEC Conditions is to avoid disputes at the end of a project by having intensive management machinery to deal with issues during the process of a project. The notification of disputes and the reference to an adjudicator is a necessary part of the detailed management philosophy under the NEC Conditions. This requires disputes to be referred to the adjudicator in a timely manner so that they can be resolved at the time. This necessarily means that for each of the stages of a compensation event, there may need to be a reference to the adjudicator to resolve the dispute.'*

[79] This is not a closed list of powers. See *Croda Europe Limited v Optimus Services Limited* [2021] WL 01535577 (2021), para. 39.

11.3.3 W2

Clause W2 must be selected where the Housing Grants Construction and Regeneration Act 1996 as amended applies.

In that context, the parties have a right to refer a dispute to adjudication '*at any time*'. Antecedent procedural steps such as requiring that the matter be first referred to party representatives cannot be imposed (but simply left as an option for the parties to take up after the dispute has arisen.). The effect of this is that in the context of clause W2 there is no timetable for referring a dispute beyond general limitation, which could mean 6 or 12 years after the events. Another feature of the legislation is the very broad definition of dispute. Attempts to restrict the type of dispute that can be referred will fail.

To reflect that statutory requirement, clause W2 does away with the table of when certain types of dispute can be referred and does not seek to impose any reference to the senior representative before adjudicating (although that is left as an option for the parties).

A dispute can be referred '*at any time*'.

What this means in practice is that a dispute about a compensation, which under clause W1 had to be raised not more than 4 weeks after the Party became aware of the action or inaction of the *Project Manager* event, is quite likely to form a subcategory of a wider dispute. The dispute referred may be whether the amount certified (including delay damages) is correct. Such a dispute may consist of many sub issues in which many compensation events and the extension of time may be in issue.

That dispute could be pursued (unless the final account provisions act as a bar) '*at any time*'.

That approach is not possible under W1 where the *WSP case* states that the categories of dispute are not overlapping but distinct and 'any other matter' cannot be used as a means to avoid the time limits for the other category of disputes.

11.3.4 The relevance of when a dispute is referred to dispute resolution

So, one may ask what is the relevance of the point in time and type of dispute that is referred to dispute resolution. The relevance is that it has an impact upon how the person or persons dealing with the dispute are likely to deal with the dispute. As we have seen, the structure of the drafting in W1 aims to produce a mechanism where the *Adjudicator* is in a sense 'stepping in' as the *Project Manager* and operating the contract as the *Project Manager* should have done. That includes the power to'

- 'review and revise any action or inaction of the *Project Manager* or *Supervisor* related to the dispute and alter a matter which has been treated as accepted or correct'.

Further

> 'If the *Adjudicator's* decision includes assessment of additional cost or delay caused to the *Contractor*, the assessment is made in the same way as a compensation event is assessed.'

If the Parties have broadly followed the contract procedure – that is, notified of a compensation event in time – and then a problem arises in, say, the rejection of the compensation event notification or the

Contractor's quotation, then the dispute is essentially resolved by operating the contract. As we have seen in the context of the compensation event drafting that requires that a forecast is made and makes an adjustment at a point in time. The guiding principle is that in the interests of certainty, contract adjustments are made at a point in time and not subject to retrospective analysis. See clause 66.3, which states that

> *'The assessment of an implemented compensation event is not revised except as stated in these conditions of contract.'*

Where there is an Accepted Programme again, we see that provides the baseline against which the delay analysis will take place.

> '63.5 A delay to the Completion Date is assessed as the length of time that, due to the compensation event, planned Completion is later than planned Completion as shown on the Accepted Programme current at the dividing date.
>
> A delay to a Key Date is assessed as the length of time that, due to the compensation event, the planned date when the Condition stated for a Key Date will be met is later than the date shown on the Accepted Programme current at the dividing date.
>
> The assessment takes into account
>
> - any delay caused by the compensation event already in the Accepted Programme and
> - events which have happened between the date of the Accepted Programme and the dividing date.'

By contrast, where there is no Accepted Programme the contract is less clear and simply states

> 'The *Project Manager* assesses the programme for the remaining work and uses it in the assessment of a compensation event if there is no Accepted Programme.'

In the context of dispute resolution what the tribunal may do and whether it will apply the prospective analysis rules or apply its own approach to adjustment of the contract is less clear cut.

In the context of adjudication, first it may depend upon what the *Adjudicator* is asked to do. If a *Contractor* asks the *Adjudicator* to decide that its programme had been incorrectly not accepted and the *Adjudicator* decides it had been incorrectly not accepted, the *Adjudicator* could decide that the programme in question is the Accepted Programme for the purpose of the contract and that may provide the basis for any subsequent dispute about an extension to the Completion Date. Where, however, that is not an option, and the *Contractor* is simply challenging the *Project Manager's* assessment, the scope for both parties to seek to challenge the analysis is potentially much wider.

11.3.5 The resistance of some tribunals to forecasting the effect of delay

Further, where the dispute is before the tribunal and the point at which the tribunal is considering the dispute is far removed from the events and the Parties have not followed the contract, it will be increasingly difficult to operate the contract without regard to information that has become known at a later date and the tribunal may consider that it would be artificial to seek to do so. In those circumstances, the analysis may be essentially a retrospective exercise based on all relevant information. See, for example, the *Northern Ireland Housing Executive* case.[80]

In the *Northern Ireland Housing Executive* case, which concerned an NEC3 PSC contract, the court had to consider what information was relevant in order to assess a compensation event. The route to assessment had been a tortuous one.

The factual; background is set out in the judgment.

'1. *The plaintiff ('the employer') awarded the defendant ('the consultant') two Asbestos Surveying Services Contracts in December 2012 for its Belfast and North East areas. The Contracts were in the form of the NEC3 Professional Services Contract (June 2005) as amended.*
2. *An instruction changing the scope of the works was communicated by the employer to the consultant at a meeting on 10 January 2013. The employer did not notify this instruction as a compensation event under the terms of the Contract but ought to have done so.*
3. *In accordance with Clause 61.1 of the contract, the employer ought to have instructed the consultant to submit a quotation on 10 January 2013 in relation to the assessment of the effects of that compensation event on the contracts for Belfast and North East.*
4. *The consultant notified that instruction as a compensation event to the employer on 21 May 2013.*
5. *Quotations were sought by the employer on 19 August 2013 (Belfast) and 22 October 2013 (North East) and provided by the consultant on 29 August 2013 (Belfast) and 31 October 2013 (North East). In accordance with Clause 63 those quotations were to assess the effects of the compensation event in each contract.*
6. *Clause 62.3 of the contract entitles the employer to reject a consultant's quotations and Clause 64.1 permits the employer to make its own assessment.*
7. *The employer rejected the consultant's quotations and assessed the effect of the compensation event as being zero on 14 November 2013 (Belfast) and 21 November 2013 (North East).*
8. *The dispute as to the effect of the compensation event was referred to Adjudication and the Adjudicator issued his temporarily binding 3 Decisions on 24 January 2014. The employer has paid the consultant in accordance with those Decisions.*
9. *The employer has issued proceedings challenging those Decisions and claims re-payment of sums paid. The consultant has also challenged those Decisions and counterclaims for further monies in addition to those awarded by the Adjudicator and paid by the employer.'*

The questions for determination by the court were agreed as follows:

1. On the true construction of the contract, and in particular Clauses 60 to 65 of the contract, is the assessment of the effect of the compensation event calculated by reference to the forecast Time Charge or the actual cost incurred by the consultant?
2. Are actual costs relevant to the assessment process in Clauses 60 to 65 of the contract?

[80] *Northern Ireland Housing Executive v Healthy Buildings (Ireland) Ltd* [2017] NIQB 43.

The case is one that very much turns on its own facts and the approach of the court in question. First, the court took a broad view as to what its function was in the context of it hearing the dispute. It was not limited to information or argument before the *Adjudicator*. There was no equivalent to clause W1.3 (7) governing adjudication, requiring the tribunal to make adjustments for additional cost or delay on the same basis as compensation events. The judge somewhat oddly took the view that the *Employer* was not 'bound' to accept a quotation, which is perhaps too simplistic as the basis for not accepting a quotation is that it is not correct. The *Employer* is in breach of contract if it simply does not accept it.

But perhaps most important is the view the judge took of the function of the court.

> *'[34] If these parties do not resolve their dispute it will fall to the court to assess the fair and reasonable "compensation" due under the contract to the consultant for the effect on it of the employer's change of instruction on 10 January 2013. This change of instructions caused the consultant, it says, increased expenditure in the period between then and the quotations it provided to the employer claiming compensation in August and October 2013 with regard to the two contracts.*
>
> *[35] Evidence, from time sheets and other material, of what the consultant actually did in that period, particularly with reference to the change in instructions, is not only relevant evidence but clearly the best evidence to assist the court in calculating the "compensation" to which the consultant is entitled. Therefore, the answer to the second question posed by the parties is yes unless the court as Tribunal is precluded from looking at the actual time charges by reason of the contract.'*

What we can see from that analysis is that the court is not approaching its task as assessing a compensation event and what follows in the Judgment is an argument as to why a 'prospective' analysis based on Time Charges is not realistic given that the assessment is not at the time the court is making it or even at the point the parties made it a 'prospective exercise'.

The judge stated at paragraph 48, as part of the discussion referring to Keating on NEC3 First Edition 2012

> *'The author considers the effect of Clause 63. I would say this with regard to Clause 63. It is clearly contemplating a situation where the employer complies with the contract and notifies the instruction at the time that he is given. He should then invite a quotation where the consultant, within two weeks, estimates the cost to it of the change of instruction. The reality, however, is that in this case that did not happen. While in the wording of the contract the word "forecast" is applicable if what should be done is done what in reality the consultant was doing in August and 12 October 2013 was making a claim for work done. It seems to me that to give an efficacious and business-like interpretation to the contract a quotation which arises in those circumstances, rather than as a genuine forecast, ought to be informed by the best information available as to the actual cost and time incurred by the consultant as a result of the instruction.'*

Whether another judge would reach the same conclusion is an open question. More recent cases have been less indulgent where the background is that a party has not done what the contract requires. But the concerns the judge has about the artificiality of a prospective analysis being done long after the events in question is a real one.

The other significant point to note is that the court was not constrained as would be an adjudicator by the wording in clause.

W2.3 (7) of the adjudication clause provides

> 'If the *Adjudicator's* decision includes assessment of additional cost or delay caused to the *Contractor*, the assessment is made in the same way as a compensation event is assessed. If the *Adjudicator's* decision changes an amount notified as due, the date on which payment of the changed amount becomes due is seven days after the date of the decision.'

In his (2006) book, *NEC3 Engineering and Construction Contract: 'A Commentary'* Brian Egglestone states with some foresight the challenge that such an approach poses for an adjudicator.

> *'This seemingly innocuous clause conceals a complex and difficult problem which, by virtue of new wording in clause 63.1 of NEC 3 will be worse than it was under ECC2.The problem arises firstly because compensation events are meant to be priced by quotations which include the contractor's risks forecasts; secondly, because the assessment process.... involves very detailed programming and costing tasks which , even if the adjudicator has the skills to undertake the task, he is unlikely to have sufficient time.'*

As explained by Egglestone, the drafting will pose a challenge for adjudicators who are constrained by jurisdictional requirements to follow the adjudication procedure and are limited to the extent to which they can take the initiative in ascertaining facts and law relevant to the dispute. The role of the *Adjudicator*, as it has developed since the advent of and development of the law under the Construction Act, is less pro-active than was perhaps first thought. The Adjudicator has to be wary of deciding matters on the basis of things that have not been put forward by the parties.

See, for example, the case of *Van Oord v Dragados* [2022] CSOH 30.

The dispute concerned delay to a harbour project. The form of contract was an NEC3 subcontract.

The following extract from the judgment sets out the difficulties that arose in the adjudication, which led to the adjudicator's decision not being enforced. (And so, the process did not provide any resolution for either party).

> *'[8] It was not in dispute that in order to assess critical delay it was first necessary to establish a baseline programme against which to assess it. That process was complicated in the present case by the fact that there was no agreed baseline programme, although numerous programmes had been prepared during the course of the sub-contract. The experts disagreed as to which of those should be used as the baseline. Mr Wilsoncroft favoured one from October 2018, whereas Mr Zucconi selected the 15 April 2019 programme as his baseline. Both had considered, but rejected, a programme prepared on 15 March 2019. Having selected their respective baselines, both experts then conducted a windows analysis of critical delay. Mr Wilsoncroft's window 3 showed the critical delay caused by CEN 048, taking 2 August 2019 as the critical date. Since Mr Zucconi had formed a different view on the facts as to the event which had caused delay, his critical dates were different: 15 June to 15 July 2019 (his window 4) and 15 July to*

> *15 August 2019 (his window 5). Neither of those showed the impact of CEN 048 since on his approach, as it was not a compensation event, it did not cause any critical delay. His analysis attributed the delay to the revetments.'*

The adjudicator's approach:

> *'[9] In the event, the adjudicator did not accept the views of either expert in their entirety. He selected the March 2019 programme as the baseline. Thereafter, having considered the evidence, he found that it was the defender's failure to progress the piling works, rather than any failure by the pursuer to complete the revetments, which had prevented the pursuer from accessing the open quay works, and which had caused critical delay. To that extent, he preferred Mr Wilsoncroft's approach to Mr Zucconi's. However, he accepted Mr Zucconi's windows analysis and delay periods in preference to those of Mr Wilsoncroft. Consequential upon his attributing the delay to the piling and not the revetments, he then necessarily had to change Mr Zucconi's allocation of liability for windows 4 to 7. The outcome of this exercise was that, for reasons which are not entirely clear, the adjudicator found that the critical date of CEN 048 was 31 July 2019 (and, further, that it extended beyond 11 October 2019, the claim for CEN 055 having been refused).*
>
> *[10] In short, the adjudicator selected as a baseline a programme which not only was not contended for by either expert, but which both experts had given reasons for rejecting; and he made an award to the pursuer based upon a critical date – 31 July 2019 – which was two days earlier than the date of 2 August 2019 proposed by the pursuer for CEN 048. Neither the date of 31 July 2019, nor the consequences of selecting it as the critical date, was canvassed with the parties. It is this which has given rise to the present controversy between the parties.'*

The defender's case was that the adjudicator was not entitled to adopt the course he did without first intimating to the parties an indication of what he had in contemplation, giving them an opportunity to address him further. Had he done so, the defender stated that it would have advanced a time bar argument based upon clause 61.3 of the subcontract.

The court agreed and refused to enforce the decision, which became a nullity.

Almost certainly the adjudicator had run out of time to reach a decision and had perhaps misunderstood the extent to which he could step in.

11.3.6 The implications of potentially complex disputes and uncertainty of outcome

Against that background, one of the key implications of failing to have an up-to-date programme is that it will complicate the compensation event procedure and may lead to real uncertainty as to how the question of extension of time will be resolved.

In that context of disputes, this suggests some broad conclusions if not definitive answers.

First, the nature of any dispute resolution process will depend in large part on the extent to which the parties have tried to follow the contract procedures. If one party has followed the procedure and the other has not it may be quite simple for an adjudicator to step in and put the process back on track.

For example, if the *Project Manager* had wrongfully not accepted the submitted programme. The *Adjudicator* would have the power to step in and review a wrongful rejection of the programme and to decide that there is an Accepted Programme.

Similarly, if the *Contractor* has not followed the process and failed to submit a compliant programme but the *Project Manager* has made a rational and timely assessment of the programme for the remaining work and used it in the assessment of a compensation event, there would be little basis for the *Adjudicator* to reject that assessment in favour of the *Contractor's* argument.

If the *Project Manager's* assessment is arbitrary or favours the *Client*, then the *Adjudicator* will have to make their own assessment. It is in this context that the contract process reaches a limit. Where both parties have consistently not followed the procedure in the contract, the suggested approach to assessment may have no application.

11.3.7 What if the process has broken down?

The contract seeks to prescribe the means by which time and money may be adjusted in clause 63.6.

> 'The rights of the *Client* and the *Contractor* to changes to the Prices, the Completion Date and the Key Dates are their only rights in respect of a compensation event.'

The contract also includes a number of deemed solutions to seek to keep the process in play where the *Project Manager* fails to operate the contract as required.

11.3.7.1 *Project Manager's* failure to apply the contract procedures

In the context of both programming and compensation events, the contract provides a mechanism to seek to keep the process alive. These mechanisms involve a warning from the *Contractor* being given and then a period for the *Project Manager* to respond and follow the process. In default of a response within the period, the *Project Manager's* response is deemed to be acceptance.

That would end the process contractually, but the *Client* has the right to ask the *Adjudicator* to decide a dispute about a matter that has been deemed to be accepted. The *Adjudicator* could it seem overturn such a deemed acceptance. However, placing the burden of a dispute on the *Client* is not something that a *Project Manager* should allow to happen. It raises several difficult issues as well as incurring greater cost than simply operating the process would involve.

11.3.7.2 General failure to follow procedures

As discussed, the contract seeks to prescribe the means by which time and money may be adjusted in clause 63.6.

However, the further the parties depart from the contract procedures, the greater potential there is to seek to evade such constraints or for a tribunal to consider broader issues in the effort to resolve the dispute.[81]

If the parties have departed sufficiently from the contract procedures, the compensation event process may have no application and the issues the *Adjudicator* or *tribunal* have to consider will be perhaps

[81] See the discussion in *Keating on NEC 2nd Edition* (Thomas and Lee, 2022: 7-182).

based on broad concepts such as 'acts of prevention' or estoppel and seeking to establish a completion date based upon general legal principles.

For the *Client*, uncertainty concerning the Completion Date is unwelcome as the *Client's* usual remedy of applying liquidated damages for delay may become unenforceable.

For the Parties, the prospect of resolving matters through formal dispute procedures is the hallmark of an unsuccessful project, placing additional strains on all participants and detracting from the focus of project delivery.

The contract gives the Parties the processes to avoid such an outcome. A collaborative and transparent approach to programming is at the heart of these processes. This is not an aspect of the contract that should be under-resourced or a process to be dabbled with.

Willis T
ISBN 978-1-83662-481-3
https://doi.org/10.1108/978-1-83662-478-320251014

Chapter 12
Conclusions and future developments

As this book has shown, the programme is at the heart of a collaborative process to use information to monitor and control delivery of construction projects. It is a subset of the total information that is processed and managed on projects. For both clients and contractors, the use of information produced in construction projects has broadened. Increasingly, clients specify information requirements and central to a client's concerns are the consistency of the underlying information and also interoperability between differing processes.

The development of information modelling has produced standards to achieve these aims. It has also become possible to use the information produced to model time not just in a linear fashion but also to use 4D information management to produce a virtual 3D model as it develops through time. Four-dimensional representations can assist in modeling complex construction sequencing or to present a picture of events.

This is not a requirement for the functions for which the programme is used in the NEC family, but it is quite possible for the client to link data produced in terms of progress records and objects that the 3D model generates.

It is also clear that the management of information must be considered at the inception of a project. One of the types of information that must considered and structured in a consistent way is the reporting of progress and the use of a common model (whether using a traditional software or as part of a broader information requirements). Clients should not rely on the supply chain to dictate the structure or content that they will provide. Although contracts may have broad requirements to collaborate and to provide information, the standard drafting does not provide the level of detail of what and how that information is to be provided to create a full structure.

It cannot be left to project managers to seek to develop such processes in detail once the contract has been entered into.

The form and content of the information required for programming purposes is understood by planners and schedulers. Often in the past planners and schedulers have been a resource that has resided in the supply chain and to some extent in consultants appointed as project managers called upon once the project is about to start. There is strong argument that planners should be involved in defining and setting the information requirements at a pre-contract stage.

Further, the increasing use of common data environments and standards for sharing data enable greater collaboration.

Information can be recorded and updated more quickly, making the information available in a time frame that allows collaboration based on better information than in the past.

The tools and the opportunity to manage projects effectively exist.[82] Those clients who manage large projects and infrastructure prescribe the relevant structures. The challenge is to then ensure they are captured and implemented for each project. The NEC family permits this across differing stages of the procurement journey of a project.

For standalone projects using the NEC4 ECC, it is incumbent on those who are appointed as *Project Managers* to be clear with the *Client* as to the demands in terms of programming. The availability of a planner or scheduler as a dedicated resource is necessary for the proper use of the contract.

For those advising on procurement, it is important that the structures for reporting progress and the format of and access to information or software should be considered at the tender stage. This is critical to the procedure for updating the programme and unless such procedures are in place it will be extremely difficult for the *Project Manager* and *Contractor* to keep an Accepted Programme in place.

Although the contract provides an 8-week period for the *Contractor* to notify of compensation events, this period can create difficulties where the timetable for assessing compensation events is out of step with the order in which they occurred. There is nothing to gain in delaying notification. Any difficulties in predicting the effect of the event can be managed in assumptions provided the *Project Manager*.

As discussed, the loss of the Accepted Programme has severe consequences for the project and much of the benefit of the risk management processes will be lost. Further, if the Parties are unable to agree adjustments to the Completion Date and must resort to a formal decision by a tribunal, there is real uncertainty as to how their dispute might be resolved. Disputes about extensions of time are notoriously difficult and costly to present or defend.

The Accepted Programme is the mechanism that allows the Parties to take control of these uncertainties alongside all of its other benefits. Every effort should be made pre and post contract to ensure that it is kept in place.

[82] Contract management software that embeds the processes into the project communications and business reporting exists such as FastDraft or CEMAR and others.

Willis T
ISBN 978-1-83662-481-3
https://doi.org/10.1108/978-1-83662-478-320251018

Table 1 NEC4 family of contracts

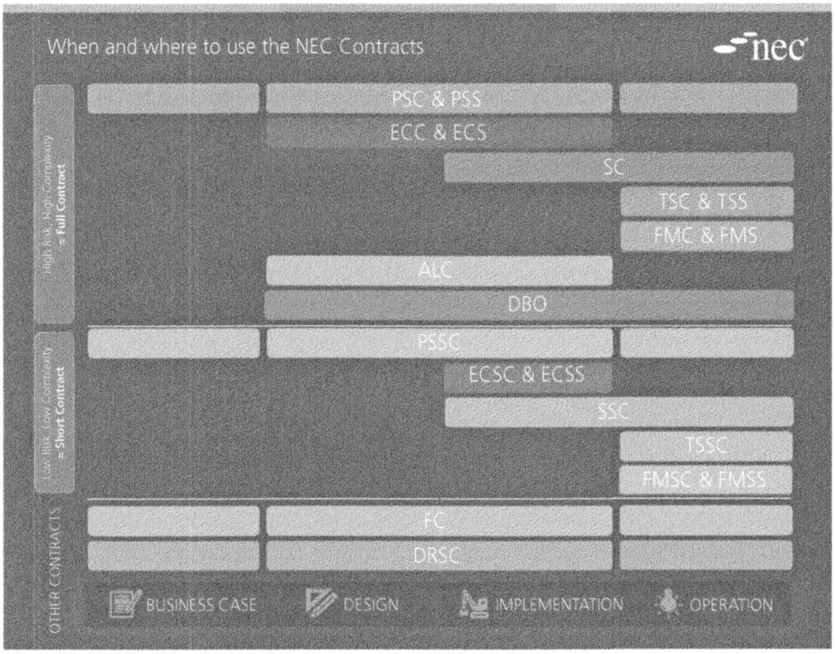

Source: © Copyright nec 2025

High Risk, High Complexity 'Full Contract' Low Risk, Low Complexity 'Short Contract'

PSC and PSS = Professional Service Contract and Professional Service Subcontract

PSSC = Professional Service Short Contract

ECC and ECS = Engineering and Construction Contract and Engineering and Construction Sub-Contract

ECSC and ECSS = Engineering and Construction Short Contract and Engineering and Construction Short Subcontract

SC = Supply Contract

SSC = Supply Short Contract

TSC and TSS = Term Service Contract and Term Service Subcontract

TSSC = Term Service Short Contract

FMC and FMS = Facilities Management Contract and Facilities Management Subcontract

FMCS and FMSS = Facilities Management Short Contract and Facilities Management Short Subcontract

ALC = The Alliance Contract

DBO = Design Build Operate Contract

FC = Framework Contract

Willis T
ISBN 978-1-83662-481-3
https://doi.org/10.1108/978-1-83662-478-320251020

Table 2 Comparison of contracts

NEC4 ECC	
Defined term	Accepted Programme
Who provides it?	*Contractor*
When	Pre or post contract
Who approves it?	*Project Manager*
Contract status	Contract Document
Failure to provide	Retention of 25% of PFWD (first programme) and/or loss of any entitlements due
Content Contractors' plans	Planned completion, float, time risk allowances Critical Path Method (CPM)
Other work	Work of *Client* or Others
Resourcing	Principal Equipment Resources
Function in relation to recording the status	Agreed record
Function of an extension of time	Adjustment of Completion Date
Obligation to update	Actual progress or planned completion change or on instruction or at interval in Contract Data

JCT 2016 SBC	
Defined term	Master Programme
Who provides it	Optional, if required, Contractor
When	Post contract
Who approves it	Architect/Contract Administrator
Contract status	None
Failure to provide	None
Content Contractors' plans	Critical Path if required by contract particulars
Other work	
Resourcing	
Function record	No
Function extension of time	Evidential if kept up to date
Obligation to update	Within 14 days of fixing of new completion date

FIDIC Second Edition 2017 (Rainbow Suite, Red, Yellow and Silver Books)	
Defined term	'initial programme' and 'revised programme'
Who provides it	Contractor
When	Post contract
Who approves it	Engineer
Contract status	Contract Document. Employer may rely upon it. Contractor must comply with it
Failure to provide	Progress reports linked to up to date programming a condition of payment
Content Contractors' plans	Planned completion, float, time risk allowances CPM
Other work	Work of employer or others
Resourcing	Method statements
Function record	Agreed record
Function extension of time	Evidential to demonstrate causation
Obligation to update	To reflect actual progress, changes or overcoming delays

ICHEME Red Book Fifth Edition 2013	
Defined term	Approved Programme
Who provides it	Contractor
When	Post contract
Who approves it	Project Manager
Contract status	Not contract document
Failure to provide	None
Content Contractors' plans	Yes
Other work	Take account of all matters that may affect completion
Resourcing	If requested by PM
Function record	Approved programme record of progress
Function extension of time	Evidential
Obligation to update	If required by PM

CIOB Time and Cost Management Contract TCM15	
Defined term	Working Schedule
Who provides it	Contractor
When	Post contract
Who approves it	Time Manager
Contract status	Contract document
Failure to provide	Correction or prevention of proceeding with work
Content	Highly detailed
Contractors' plans	
Other work	Take account of all matters that may affect due dates
Resourcing	If requested by PM
Function record	Approved programme record of progress
Function extension of time	Evidential – conclusive evidence
Obligation to update	If required by Time Manager

Willis T
ISBN 978-1-83662-481-3
https://doi.org/10.1108/978-1-83662-478-320251019

Table 3 NEC4 ECC – Practice Note 1.1

'NEC4 ECC – PRACTICE NOTE 1.1 Assessing delays due to compensation events

This practice note has been prepared for the NEC4 Engineering and Construction Contract, as amended in January 2019. It is an updated version of the practice note originally issued following the publication of NEC4 in 2017. This update takes into account the changed drafting of clause 63.5, which clarifies how the contract is managed. The approach set out in this note should be followed on other NEC4 contracts.

The key requirements for assessing delays to Completion Dates and Key Dates due to a compensation event under ECC clause 63.5 are to determine the effect that the compensation event has upon planned completions and

- use the Accepted Programme current at the dividing date to do so. The objective is to assess the effect due only to the compensation event and not due to other events. The assessment takes into account;
- any delay caused by the compensation event already in the Accepted Programme and
- events which have happened between the date of the Accepted Programme and the dividing date.

In some cases, the effect of a compensation event may have already impacted on the Accepted Programme for example where an event was raised as an early warning (clause 15.1) and then later identified as a prevention event (clause 19). In such cases any delay already reflected in the Accepted Programme needs to be taken into account when assessing the impact of the compensation event on planned Completion.

In terms of events that may be taken into account between the date of the Accepted Programme and the dividing date these could be other compensation events that have occurred prior to the dividing date and any delays resulting from events which are not compensation events.

In practical terms, the parties need to decide which activities on the Accepted Programme each compensation event affects, and what the delay impact is to each activity. The durations of activities which are not completed at the dividing date are changed where appropriate to recognise the effect of the compensation event. Activities which have been completed may have been affected by the compensation event, and if so, the time impact must be recognised. However, the actual timing of completed work is not further adjusted.

In assessing the effect, risk allowances and mitigation measures should be included on the basis that the *Contractor* reacts competently and promptly to the event. The intention is that the logic in the Accepted Programme is used as the basis for determining the effect of the compensation event upon planned completions, subject only to changes required as a direct result of the compensation event assessment.

If helpful, users may produce a version of the programme specifically for the purpose of the compensation event assessment. This can be used to support and explain the assessment. However, it is important to understand this programme is not a programme submitted for acceptance but is used only for assessing the compensation event.

If the Accepted Programme at the dividing date is a long way out of date, the approach to be taken depends upon the reasons for the out-of-date programme.

- If the parties have set long intervals between submissions of revised programmes in Contract Data, then the practical guidance above should be followed.
- If the *Contractor* has not submitted a revised programme for acceptance in accordance with clause 32.2, then the parties should seek to agree a programme to be used for the assessment, failing which the *Project Manager* should make an assessment in accordance with clause 64.
- If a recent programme submitted by the *Contractor* has not been accepted by the *Project Manager*, then the parties should seek to agree a programme to be used for the assessment. If the Parties agree that the reasons given for non-acceptance do not affect the assessment of the compensation event, then the submitted programme should be used to identify the changes. If a programme to be used for the assessment cannot be agreed the *Project Manager* must make an assessment under clause 64, based on the previously Accepted Programme.

It may be that the assessment of delay using the Accepted Programme identifies an obvious error which does not show a genuine forecast effect of the compensation event on planned completions. In this situation the *Project Manager* and *Contractor* should resolve the problem in the logic of the Accepted Programme. To recap, the objective is to assess the effect due to the compensation event and not due to other factors.

If multiple compensation events arise the assessments should be carried out sequentially, separately and prospectively. However, when a number of events arise over a short period of time, it may be sensible, by agreement, to consider them together. The impact of other compensation events existing prior to the dividing date should be recognised before making the new compensation event assessment.'

Willis T
ISBN 978-1-83662-481-3
https://doi.org/10.1108/978-1-83662-478-320251015

Table 4 Table of references and contracts

CIOB (2018) *Guide to Good Practice in the Management of Time in Major Projects*, 2nd edition. Wiley, Chichester, UK.

CIOB (2021) *CIOB Planning Protocol 2021*. CIOB, Bracknell, UK.

Eggelston B (2006) *NEC3 Engineering and Construction Contract*, 2nd edition. Blackwell, Oxford, UK.

Gibson R (2007) *Construction Delays Extensions of Time and Prolongation Claims*. Taylor Francis, Oxford, UK.

Gibson R (2015) *Practical Guide to Disruption and Productivity Loss on Construction and Engineering Projects*. Wiley, Chichester, UK.

Hughes K (2022) *NEC 4 A Guide for Better Delivery of Projects in the Building Sector*. ICE Publishing, London, UK.

International Organisation for Standardisation (2017) ISO 44001. Collaborative business relationship management systems-requirements and framework. ISO, Geneva, Switzerland. www.iso.org.

International Organisation for Standardisation (2018) ISO 19650-1:2018. Organization and digitization of information about buildings and civil engineering works, including building information modelling-information management using building information modelling. ISO, Geneva, Switzerland. www.iso.org.

Latham M (1994) *Constructing the Team – Final Report of the Joint Government/Industry Review of Procurement and Contractual Arrangements on the United Kingdom Construction Industry*. HMSO, London, UK.

Mosey D (2019) *Collaborative Construction Procurement and Improved Value*. Wiley, Chichester, UK.

NEC4 User Guide (June 2017) *Managing an Engineering and Construction Contract, Volume 4*. https://www.neccontract.com/products/contracts/nec4/user-guides-and-flow-charts/nec4-managing-anengineering-and-construction-contract

NEC4 Users Guide (June 2017, with amendments October 2020) *Preparing an Engineering and Construction Contract, Volume 2*. Thomas Telford Limited, UK. https://www.neccontract.com/products/contracts/nec4/user-guides-and-flow-charts/nec4-preparing-an-engineering-and-construction-contract#:~:text=For%20users%20seeking%20to%20prepare%20their

Society of Construction Law (2017) Society of Construction Law Delay and Disruption Protocol. SCL, Leicestershire, UK.

Thomas D (2012) Keating on NEC3, 1st edition. Sweet and Maxwell, London, UK.

Thomas D and Lee K (2022) *Keating on NEC*, 2nd edition. Sweet and Maxwell, London, UK.

Contract management software

CEMAR https://thinkproject.com/products/cemar/

FastDraft https://www.builtintelligence.com/pages/fastdraft-construction-contractmanagement-system-nec3-nec4-jct-fidic

Contracts

CIOB Time and Cost Management Conditions TCM15. *Chartered Institute of Building*, Bracknell, UK. www.ciob.org

FIDIC Second Edition (Rainbow Suite, Red, Yellow and Silver Book). *The International Federation of Consulting Engineers*. FIDIC, Geneva, Switzerland. www.fidic-org/bookshop

JCT 2016 Editions. *Joint Contracts Tribunal*. JCT, London, UK. www.jctltd.co.uk

Institution of Chemical Engineers (2013) *IChemE Red Book Fifth Edition*. Institution of Chemical Engineers. IChemE, Rugby, UK. www.icheme.myshopify.com

NEC Contracts. Thomas Telford Limited, UK. www.neccontract.com

NEC4 Users Guide, Preparing an Engineering and Construction Contract Volume 2 (June 2017, with amendments October 2020). Thomas Telford Limited, UK. https://www.neccontract.com/products/contracts/nec4/user-guides-and-flow-charts/nec4-preparing-an-engineering-and-construction-contract#:~:text=For%20users%20seeking%20to%20prepare%20their

NEC4: Effective Preparation and Management of the Construction Programme

Willis T
ISBN 978-1-83662-481-3
https://doi.org/10.1108/978-1-83662-478-320251016

Index

Printed and bound by CPI Group (UK) Ltd, Croydon, CR0 4YY
21/06/2026
14903998-0002